Counting the Eons

SCIENCE ESSAY COLLECTIONS BY ISAAC ASIMOV

From Fantasy and Science Fiction

Fact and Fancy
View from a Height
Adding a Dimension
Of Time and Space and Other Things
From Earth to Heaven
Science, Numbers and I
The Solar System and Back
The Stars in Their Courses
The Left Hand of the Electron
The Tragedy of the Moon
Asimov on Astronomy
Asimov on Chemistry
Of Matters Great and Small
Asimov on Physics
The Planet That Wasn't
Asimov on Numbers
Quasar, Quasar, Burning Bright
The Road to Infinity
The Sun Shines Bright
Counting the Eons

From Other Sources

Only a Trillion
Is Anyone There?
Today and Tomorrow and—
Science Past—Science Future
The Beginning and the End
Life and Time

Counting the Eons

★

by Isaac Asimov

Doubleday & Company, Inc., Garden City, New York

1983

The following essays in this volume are reprinted from *The Magazine of Fantasy and Science Fiction*, having appeared in the indicated issues:

Milton! Thou Shouldst Be Living at This Hour (*August 1980*)
Getting Down to Basics (*September 1980*)
The Word I Invented (*October 1980*)
Counting the Eons (*November 1980*)
Beyond Earth's Eons (*December 1980*)
All and Nothing (*January 1981*)
Nothing and All (*February 1981*)
Light as Air? (*March 1981*)
Too Deep for Me (*April 1981*)
Under Pressure (*May 1981*)
Yes! With a Bang! (*June 1981*)
The Runaway Star (*July 1981*)
The Dance of the Stars (*August 1981*)
And After Many a Summer Dies the Proton (*September 1981*)
Let Me Count the Days (*October 1981*)
The Crucial Asymmetry (*November 1981*)
Let Einstein Be! (*December 1981*)

Library of Congress Cataloging in Publication Data

Asimov, Isaac, 1920–
Counting the eons.

1. Astronomy—Addresses, essays, lectures. I. Title.
QB51.A785 520 82–45068
ISBN 0-385-17976-6
AACR2
Library of Congress Catalog Card Number 82–45068

Printed in the United States of America

To Hugh O'Neill and Betty Prashker

despite the contracts and big advances they force on me

CONTENTS

INTRODUCTION

When I first started writing these essays in *The Magazine of Fantasy and Science Fiction* back in 1958, I was unaware of a number of things.

First, I was unaware when I first started what a long-term undertaking it was to be. I've been going for nearly a quarter of a century now without missing an issue and I'm told firmly by the Gentle Editor of the magazine, Edward L. Ferman, that my job is not for life but for eternity so that I had better not think of squeezing out of the agreement by anything as sleazy as dying. I think that's rather cruel of him but since he threatens to sue me for everything I have down to the gold in my teeth, I had better stay alive.

Second, I was unaware when I first started that the essays would ever reach anyone but the steady readers of the magazine, or that each individual essay would be in circulation for more than one month. Beginning in 1962, however, the pleasant people at Doubleday & Company, Inc., have been publishing collections of the essays as steadily as I have been producing them and they, too, have gotten it across to me verbally that mortality is not an option. The result is that these essays cannot be dismissed as ephemera only, to be recovered only from the collections of hard-line science fiction fans, or from the jumble in bookstalls devoted to old and decaying magazines. They have reached more or less permanent form in hard covers and soft covers and I am therefore forced to take them more seriously. (They earn me more money than I would have thought possible when I first started, come to think of it—but I care nothing for money. Well, almost nothing.)

Finally, I was unaware that I had any mission at all in these essays when I first started. I was writing only for my own amusement and, perhaps, for the amusement of those of the readers of the magazine who were interested enough to read them.

But now I have a mission!

The self-styled Moral Majority is loose in the land. They are "literalists," believing in the letter of the Bible as they understand that letter to be. They don't let it go at that, but are determined that everyone else is going to believe the letter of the Bible as *they,* the "Moral Majority" understand that letter to be.

Now don't get me wrong. I have no objection to faith and belief. I have faith and belief myself. I believe that the universe is comprehensible within the bounds of natural law and that the human brain can discover those natural laws and comprehend the universe. I believe that nothing beyond those natural laws is needed.

I have no evidence for this. It is simply what I have faith in and what I believe.

And if I arrogate to myself the right to have faith and belief, I must, in good conscience, grant others the same right.

Nor do I quarrel with the attempts on the part of those who differ from me in matters of faith; of those, for instance, who believe in the letter of the Bible; to attempt to persuade others to believe as they do.

It is my profession after all, as exemplified in these essays, to persuade others to believe as I do and to accept my notions of scientific evidence and rational thought. And I cannot do this in good conscience unless I am willing to grant those who disagree with me the same chance to put their persuasive faculties to work.

However, the self-styled Moral Majority is not content to believe and not content to persuade. They rally their people behind them—simple, unsophisticated people for the most part—and are attempting to censor books, movies, television and so on by united action; to cow and terrify schoolteachers and librarians; to bullyrag legislators; and to call the power of the law and the state to *enforce* their views on the public and to erase all freedom of thought.

Specifically, for instance, they are the driving force behind "scientific creationism" (as they call it) against the fact of evolution. The creationists believe that the universe was created not more than ten thousand years ago by the agency of a "Creator," and that all the species of animals were separately created.

They have no evidence for this, none at all.

What the creationists *call* evidence is to point at petty disagreements among thoughtful paleontologists concerning some of the details of the evolutionary mechanism (although, of course, those same paleontologists all accept the fact of evolution itself).

What the creationists *call* evidence is various distortions of scientific concepts, distortions that no bright high school student would make.

What the creationists *call* evidence is quotations from various other creationists whom they call "scientists" on the basis of such things as a degree in engineering; making use of no quotes, moreover, that ever offer any evidence worthy of the name.

Given all this, the creationists are nevertheless driving the teaching of evolution out of the public schools by conducting a reign of terror against it. They are attempting to have laws passed that will *force* the teaching of creationism. They are trying to determine what is scientifically valid by legal fiat, and once *that* principle is set, what happens to our liberties? Why may not the various legislatures then compel the teaching of the flat earth, the Easter bunny and the tooth fairy—and, of course, that good old stork as baby agent?

I decided some time ago, therefore, that since I have a platform, I will not hesitate to use it to support, specifically, those areas of science that the creationists most oppose.

A number of the essays in this book, for instance, are not in accord with the Babylonian science which creationists accept. One of those essays is "Counting the Eons," which I chose for the title of this book as a whole.

Why not? I don't believe Babylonian science. Twenty-five hundred years have passed and we have learned a little since the Babylonians—even if the creationists haven't.

If the creationists had their way, this book and many others would be burned, and we would all be compressed into the narrow, narrow bounds of their tiny and unthinking view of the universe.

Well, I, for one, refuse to cower before them, refuse to truckle to them, refuse to compromise with them, and intend only to fight them —in order to preserve my simple right to think.

I wouldn't dream of forcing you to join me in this fight; but I would hope to persuade you to do so by the contents of this book and of others that I write.

A ★ THE EARTH

1★ LIGHT AS AIR?

Those of you who follow this essay series know that I am fascinated by coincidence.

I don't attach any supernatural importance to it; I know it is inevitable and that lack of coincidence would be more surprising than any coincidence. And yet, when they occur . . .

Not very long ago, my wife, Janet, and I were walking down a neighborhood street, having just eaten at a local restaurant, and Janet casually pointed out a small shop named Levana, where she bought Middle Eastern bread and pastries.

She pronounced it as spelled, and I corrected her and gave it the proper Lithuanian Yiddish pronunciation. I then went on to explain that it was the Hebrew word for *moon,* because I know that she enjoys having me inform her of all sorts of miscellany that I dredge up from somewhere within me.*

A distant memory stirred of the days, four decades back, when I used to attend Yiddish musicals that featured a hilarious comic named Menasha Skulnik.

"In fact," I said, "one of the most famous Yiddish popular songs of my childhood featured the word."

Despite the fact that forty years had passed during which I had not once, to my conscious knowledge, heard the song or thought of it, I proceeded to sing the first line, "Shane vie die levana" ("Beautiful as the moon"), with the correct tune, I assure you.

* At least I *think* she enjoys it.

Two minutes after that, we reached a secondhand bookstore and since Janet is constitutionally incapable of passing a secondhand bookstore without walking in, she entered, and I, of course, followed.

Two minutes after we had entered, the proprietor (who also sold secondhand records) put on a record for some unknown reason, and you know perfectly well what it was. It was Menasha Skulnik singing "Shane vie die levana."

Intellectually, as I told you, I understand coincidences and don't allow myself to be impressed by them. Emotionally? Well, that's another thing altogether.

I became totally incoherent. I jumped up and down. I pointed to the record player and finally managed to gasp out, "That's the song! That's the song!" to a totally flabbergasted Janet, who thought I was having a fit.

And it was indeed a fit of a sort. I could feel my ears buzzing, my vision darkening, and a fearful internal pressure. I knew my blood pressure was hitting the ceiling and that that silly little concatenation of events was bringing me within measurable distance of an apoplectic stroke.

So I forced myself to stand still, close my eyes, take a deep breath, and think of something neutral—and the danger passed.

But the thought of my blood pressure led me on to think of pressure in general, and that brings me to the subject of the present essay.

The pressure to which we are most accustomed and to which we are all constantly subjected is the pressure of the air upon ourselves. The atmosphere, having mass, is attracted by the earth's gravitational field and therefore has weight. Consequently, it weighs down and presses upon us just as surely as a lump of iron would.

Nor does it merely press downward. Gases (and liquids, too) flow, so that the pressure is transmitted in all directions, sideways and upward, as well as downward.

We don't feel the pressure of the weight of the atmosphere, because the liquid contents of the body's cells exert the same pressure in every direction. The atmosphere presses inward on our body and the body's liquid contents press outward; the two balance and we feel nothing.

The pressure of the atmosphere varies from time to time, due to the effect of temperature, of height above sea level, of air movement.

The internal pressure of the body varies to suit, but not always rapidly enough. When the pressure changes more rapidly than usual, some of us have feelings of malaise.

The Eustachian tube is too narrow to allow a rapid equalization of air between the middle ear and the outside, so when we change elevation rapidly, as in an elevator, we have to swallow to let our ears "pop."

We can, however, specify standard conditions. We can suppose that we are dealing with quiet air at 0° C. at sea level. In that case, what is the air pressure?

The simplest answer is "1 atmosphere," where an "atmosphere" is defined as the pressure of earth's air blanket under standard conditions.

This sounds as though we were arguing in circles, but it has its useful aspects. If I told you that the air pressure on Venus was about 90 atmospheres and that on Mars it was about 0.01 atmosphere, you would have something that was informative.

Nevertheless, to say that the pressure of our atmosphere is 1 atmosphere does seem to leave us a little short of content. Can we do better?

In ancient times, no thought was given to the fact that air might have weight and exert pressure. The sensation of weight was absent and it seemed reasonable to accept that as evidence that the reality of weight was absent as well. Therefore we still have "light as air" as a familiar cliché, and are perfectly capable of speaking of "airy nothings."

It was noticed at various times, however, that no matter how well built a pump and no matter how assiduously people worked the pump handle, water could never be raised more than about 34 feet above its natural level.

The reason for this is that a water pump pulls some of the air out of the pump's interior. The air pressure on the water surface outside is thus greater than the air pressure within the pump, and that excess air pressure outside pushes water up the pump cylinder to a height where the water pressure inside plus the diminished air pressure balances the total air pressure outside.

By the time the water reaches a level 34 feet above its natural level, however, it exerts a pressure all by itself that is equal to the external air pressure. Nothing more can be done; the water can be pumped no higher since there exists no additional air pressure outside to do the job.

Careful measurements at 4° C. (when water is at maximum density) shows that the weight of a column of water 33.899 feet high produces a pressure that exactly balances air pressure. We can therefore say that:

1 atmosphere = 33.899 feet of water
or 11.230 yards of water
or 406.79 inches of water

This view that air had weight and exerted pressure was compellingly demonstrated in 1643 by the Italian physicist Evangelista Torricelli (1608–47), who substituted mercury for water. He filled a four-foot-long tube (closed at one end, of course) with mercury, and tipped it into a trough of mercury. Some mercury ran out, but only some. A column of mercury was left standing, held up by the air pressure outside.

Since mercury at 0° C. is 13.5951 times as dense as water, a column of mercury of given cross section will weigh as much as a column of water of the same cross section that is 13.5951 times as high as the mercury column. This means that if air pressure balances 33.899 feet of water, it will balance 33.899/13.5951, or 2.493 feet, of mercury. Torricelli did indeed find that the mercury column stood at the height it was supposed to as closely as he could measure it. We can therefore say that:

1 atmosphere = 2.493 feet of mercury
or 29.92 inches of mercury

Torricelli did more than demonstrate that air has weight and exerts pressure. He did more than explain why water (or any liquid) can only be pumped so high. When some of the mercury poured out of Torricelli's tube, it left a vacuum behind, the first good vacuum ever produced by human beings. A vacuum produced in this manner is therefore called a *Torricellian vacuum.* As if that weren't enough, Torricelli had, by means of his ingenious demonstration, invented the mercury barometer, an instrument which, to this day, varies from Torricelli's tube only in detail.

By using a column of mercury and measuring its height from time to time, we can detect changes in the "barometric pressure" of the atmosphere.

As weather conditions vary, as temperature goes up or down, as stormy weather with rapid air movement succeeds calm weather and vice versa, as the barometer itself is moved from place to place, the measured air pressure changes to a minor degree. Since such changes (experience tells us) mark whether the weather will change, and how, the barometric pressure and the direction of its change, whether up or down, is an invariable accompaniment of the weather report.

The barometric pressure never varies very much, of course. The highest recorded value at sea level was not quite 32 inches, and the lowest just under 26 inches. Except under exceptional circumstances, however, the barometric readings stay within less than an inch of 30 inches of mercury.

To be sure, the inch is an outmoded unit of length. The world (with the exception of a few nations of which the two most powerful are the United States and Liberia, in that order) uses the metric system. A unit of length in the metric system is the *meter,* which is equal to 39.37 inches. A *centimeter* is $\frac{1}{100}$ of a meter, or 0.3937 inch (or about $\frac{2}{5}$ of an inch). A *millimeter* is $\frac{1}{1000}$ of a meter, or 0.03937 inch (or about $\frac{1}{25}$ of an inch).

Since 29.92 inches equals 75.97 centimeters, we can say:

1 atmosphere = 0.7597 meter of mercury
or 75.97 centimeters of mercury
or 759.7 millimeters of mercury

Usually, of course, the measurements are not given their fullest precision. If asked what the value of 1 atmosphere is, the usual answer would be "30 inches of mercury" in the United States, and "760 millimeters of mercury" everywhere else.

So far, though, I've only been comparing weights, and matching up the pressure of a column of water or of mercury with a column of air. There is, however, an important difference between weight and pressure.

Suppose we had a cylinder that was 1 square inch in cross section. If it contained mercury to the height of 30 inches, there would be a certain weight of mercury resting on the square inch at the bottom of the cylinder.

Suppose next that we had a cylinder that was 4 square inches in cross section. If it contained mercury to the height of 30 inches, there would be four times the weight of mercury in this cylinder as in the smaller one; but that quadrupled weight would be resting on an area four times as large as in the previous case. The weight on each square inch would be the same in the two cases.

Once we define pressure as "weight per unit area," then the cross section of the tube doesn't matter; only the height.

If we wish to describe 1 atmosphere in terms that are immediate and dramatic, then, we should ask: "What is the weight of a column of air that is resting upon one square inch of our bodies?" It would equal the weight of a column of mercury 29.92 inches high and 1 square inch in cross section. The weight of such a column of mercury is 14.696 pounds and therefore:

1 atmosphere = 14.696 pounds per square inch

This is invariably startling to anyone who first comes across the fact. Mark out a square inch on your arm, or on any part of your body, for that matter, even the most delicate; and there is a weight of 14.696 pounds of air pressing down upon it (or sideways, or up —for air pressure is exerted in all directions equally).

The total surface area of a rather pudgy person of average height (me, for instance) is about 2,950 square inches. That means I bear upon my body a total weight of 21.7 *tons* of air.

Yet I move about freely.

Partly, this is because the weight is evenly spread out over me. What counts is not the total weight, but the weight per unit area— that is, the pressure. Even more important, the internal pressure of my fluid contents (as I said earlier in the essay) exactly balances the external air pressure.

The crucial nature of the difference between weight and pressure can be shown by a thought experiment. Imagine an ounce weight resting upon your forearm. Nothing much happens. Now imagine a needle on your skin, point-downward, and an ounce weight balanced on the top end of the needle. The needle will puncture your skin.

In the first case, the ounce weight is distributed over a considerable area of your skin, and the weight pressing down on any region as small as a needlepoint is very tiny. With the needle between the weight and the skin, all the downward push of the weight is concentrated on the tiny area of the needle's point. The pressure at that point is enormous and the needle is forced through the skin.

Again, when you drive a nail into wood, you place the pointed end on the wood and hammer the flat end, and in it goes easily. If you placed the flat end on the wood, no amount of hammering the pointed end would do you any good. Even more dramatically, you place your thumb on the flat end of a thumbtack and the pointed end on the wood and press. The feeble push of your thumb is enough to drive the pointed end into the wood because the pressure transmitted through the point is enormous.

Pounds and inches are, however, passé.

Since a centimeter is 0.3937 inch, a square centimeter is 0.3937 × 0.3937, or 0.1550 square inch (or about $\frac{2}{13}$ of a square inch).

The metric unit of mass (usually used also for weight) is the gram, which is equal to 0.00220462 pound (or, roughly, $\frac{1}{450}$ of a pound). Therefore 1 pound per square inch equals 70.307 grams per square centimeter. Multiply that by 14.696 and you have:

$$1 \text{ atmosphere} = 1{,}033.2 \text{ grams per square centimeter}$$

It is also possible to use kilograms as the unit of mass (or weight) and square meters as the unit of area. A kilogram is equal to 1,000 grams. A meter is equal to 100 centimeters so that a square meter is equal to 100 × 100, or 10,000 square centimeters. Therefore, 1 kilogram per square meter is equal to 0.1 gram per square centimeter. It follows that:

$$1 \text{ atmosphere} = 10{,}332 \text{ kilograms per square meter}$$

Pounds and grams and kilograms are units of mass, and although they are casually used as units of weight as well (even by scientists) it is wrong to do so. Weight is proportional to mass, but weight is *not* mass.

The sensation of weight is the result of a response of earth's gravitational field. The atmosphere is attracted to the earth and pushes down against the earth's surface thanks to its interaction with earth's gravitational field. This *push* is what gives rise to the sensation of weight, and a push is a *force*. If we are dealing with pressure as a weight per unit area, we should properly seek to make use of units of force.

Using units of mass for weight is easily understandable and will solve our problems adequately as long as we remain at earth's sea level and deal with gravitational interactions, such as columns of air, water, or mercury. If, on the other hand, we deal with something such as the hammering of a nail or the pressing of a thumbtack, we find we can no longer make sense out of weight per unit area. After all, as you wield the hammer harder and harder, its weight doesn't change. The force it exerts does change. We *must,* then, use force per unit area.

To work out units of force, consider that a force is a push or a pull that brings about an acceleration. In fact, it is the presence of an acceleration that demonstrates the existence of a force.

Suppose there were a force capable of causing a mass of 1 gram to accelerate at a rate of 1 centimeter per second per second. That is, you start with a mass of 1 gram at rest and therefore moving at 0 centimeters per second. The exertion of the force would mean that 1 second later the mass is traveling at a speed of 1 centimeter per second in the direction the force is pushing. Another second later and it is moving 2 centimeters per second. Yet another second later and it is moving 3 centimeters per second, and so on.

Such a force has a magnitude of "1 gram-centimeter per second per second." Scientists, even as you and I, however, would find it boring to be continually repeating "gram-centimeter per second per second" and they replace it with the syllable *dyne* from a Greek word meaning "force."

The unit of force then is the dyne, and 1 dyne is capable of accelerating 1 gram at a rate of 1 centimeter per second per second. A force of 2 dynes would accelerate 2 grams at a rate of 1 centimeter per second per second, or 1 gram at a rate of 2 centimeters per second per second, and so on.

One gram of weight pressing down upon a square centimeter exerts a force of 980.68 dynes upon that square centimeter. Since an atmosphere is equal to 1,032.2 grams per square centimeter, we can also say that:

1 atmosphere = 1,013,200 dynes per square centimeter

As long as we deal in dynes per square centimeter, we can handle the pressure on a nail, or that of rocket exhausts, in the same way we deal with pressures arising from weight.

The only trouble is that *dynes per square centimeter* has seven syllables and that the number 1,013,200 is large. People are continually dealing with air pressure and the use of seven digits and seven syllables can be wearying after a while. Fortunately, we can always invent a term and define it appropriately.

Scientists define a *bar* (from the Greek word for *heavy*) as 1,000,000 dynes per square centimeter. That means that:

1 atmosphere = 1.0132 bars

This is very convenient since if you want to make a quick rough estimate that involves atmospheric pressure, you can so arrange matters as to let it equal 1 bar and that simplifies the arithmetic enormously.

Of course, many pressures that are dealt with are much smaller than atmospheric pressure, and for those purposes you can use the *millibar,* which is equal to $1/1000$ of a bar. You can therefore say:

$$1 \text{ atmosphere} = 1{,}013.2 \text{ millibars}$$

In recent years, there has been a move to get some order out of the metric system. After all, you can measure pressure as grams per square centimeter, kilograms per square decimeter, hectograms per square meter, milligrams per square kilometer, and so on. You can measure force as gram-centimeters per second per second; kilogram-meters per minute per minute; milligram-decimeters per hour per hour; and so on.

These all involve metric units; all are equally valid and useful under the proper circumstances. Still, if different people use different combinations of metric units, there is always the necessity of converting one into the other in order to compare results and observations. These conversions are pitfalls of possible arithmetical error, for they involve the shifting of decimal points, and everyone who has ever tried to shift one under stress knows that they are just as likely to move in the wrong direction as in the right one.

There has now, however, come into use something called, in French (the international language of measure, since it was the French who invented the metric system), the Système International d'Unites, which is International System of Units in English. In brief it is referred to as the SI.

In the SI, users are restricted to specific units for the various types of measurement; the meter, for instance, for length, the kilogram for mass, and the second for time.

A dyne is 1 gram-centimeter per second per second, but neither gram nor centimeter is basic in the SI scheme of things. The dyne is therefore not an SI unit and should not be used.

The unit of force in the SI system is 1 kilogram-meter per second per second. Since a kilogram is equal to 1,000 grams and a meter is equal to 100 centimeters, a kilogram-meter is equal to $1{,}000 \times 100$, or 100,000 gram-centimeters. Hence 1 kilogram-meter per second per second is equal to 100,000 gram-centimeters per second per second; that is, to 100,000 dynes.

For simplicity, 1 kilogram-meter per second per second is defined as 1 newton, after Isaac Newton (1642–1727), who first defined a force in terms of acceleration. Therefore, 1 newton equals 100,000 dynes.

I earlier gave the measurement of pressure in terms of dynes per square centimeter, but the centimeter is not a fundamental SI unit, either. We have to use the square meter and measure pressure in terms of newtons per square meter. Since a newton is equal to 100,000 dynes and a square meter is equal to 10,000 square centimeters, 1 newton per square meter is equal to 10 dynes per square centimeter.

In SI units, then:

1 atmosphere = 101,320 newtons per square meter

To speak of newtons per square meter is to make use of six syllables. Therefore 1 newton per square meter is referred to as 1 pascal, after the French mathematician and physicist, Blaise Pascal (1623–62) who, in 1646, demonstrated that barometric pressure decreased as one went up a mountainside. (Actually, Pascal didn't go up the mountainside himself. He sent his brother-in-law, with two barometers, scrambling up there. But, then, what are brothers-in-law for?) Consequently,

1 atmosphere = 101,320 pascals

There you are, then. I have given you air pressure in terms of every type of unit I could dredge up. All are equally valid and each has its conveniences.

Yet it doesn't exhaust the subject since there are other pressures than those of the atmosphere. We'll go into that in the next chapter.

2 ★ TOO DEEP FOR ME

You can't always tell what hidden talents you might have. I discovered one that I had when I was at sea recently—for though I will not fly, I don't mind ships. In fact, I like ships.

As someone who is, in general, a nontraveler, I always assume that the hardships of travel, whatever they might be, would be sure to lay me low. Therefore, I was rather apprehensive about what would happen when I encountered a rough sea. To be sure, in over a dozen cruises in the course of eight years I had not encountered any but I knew that the day was bound to come.

And then, last month, on the way back from Bermuda, the ship began to behave like an irritable bronco. At least half the passengers decided that this was the moment they were waiting for to spend a lot of time in deep thought, and they retired to their cabins to meditate. I, however, discovered with delight that the motion didn't bother me very much.

It would perhaps be less than truthful to say that I was in absolutely top form, but I ate freely and moved through the corridors (reeling somewhat from side to side as the ship rolled) lightheartedly. In fact, so jolly was I at discovering my relative immunity to rough seas that at one point I burst into song.

A ship's officer, passing by, stopped and said to me, "You sound happy."

"I *am* happy," I said.

Whereupon he said, "What an inspiration you should be to the other passengers."

"No, no," I said, nervously, "don't tell them. If they find out I've been singing, they'll kill me."

So let's talk about the ocean.

In the previous chapter, we discussed air pressure in all its ramifications, and one of the ways to express the amount of pressure we feel at the bottom of the atmosphere, at sea level, is to say that standard atmospheric pressure can support a column of water 33.899 feet high. Or, to use the metric system, a column of water 10.332 meters high.

That means that if you* were to dive into the sea to a depth of 10.332 meters, you would feel the effects of two atmospheres of pressure. One of them would be due to the atmosphere itself, which would transmit its effect through the water, and the other would be due to the water.

The additional atmosphere of pressure can be withstood since pearl divers routinely dive as deep as 15 meters below the surface of the sea and do so without any sort of protection. At such a depth,

* *You,* not I. I'm a creature of the surface. Any depth is too deep for me.

they experience 2.5 atmospheres of pressure and, if all goes well while they're down there, they return to the surface in fine shape.

For every additional 10.332 meters of depth below the surface of the ocean, however, there is an additional atmosphere of pressure, and that pressure, as it builds up, eventually becomes too much to handle.

In adventure stories, for instance, heroes have been known to attempt to escape pursuit by ducking underwater in a pond or lake amid a clump of reeds. Such a hero can't stay under very long, of course, since he has to breathe, but our hero is a man of infinite resource. He breaks off a reed, which is hollow, tears off the top, puts one end in his mouth and allows the other to extend just above the water level, where it won't be noticed among the other reeds.

He can then remain underwater for a long time, breathing in this unobtrusive manner—and, in fact, in all the adventure stories I've ever read or viewed, in which the hero uses this ruse, he succeeds in escaping.

As a matter of fact, alligators and hippopotamuses make use of a trick much like this. Their nostrils are placed in bulges at the top of their snouts. They can remain just about completely submerged in water with only their nostrils above water.

But how far can we go in this fashion? Can we prowl around the bottom of lakes in comfort simply by making use of a nice wide plastic tube long enough to reach the surface—and breathing through it?

No, we can't go too far with this sort of thing. First, the longer and, therefore, the more voluminous the tube, the less likely we are to be breathing fresh air.

Suppose you suck in a deep breath through the tube and fill your lungs. Next you breathe out. Some of the air you exhale doesn't reach the top of the tube. In fact, the tube is filled with exhaled air. When you breathe in, fresh air will enter the tube, but it can't enter the lungs till you have inhaled all the exhaled air that was in the tube.

If the tube is long enough and wide enough, *all* the exhaled air remains in that long tube, and it is all you will get when you inhale again. You will breathe the same air over and over again and it will not be long before you suffocate. If you make the tube thin, to keep the volume down, you won't be able to suck air in fast enough and you will suffocate anyway.

There is a second objection to the breathing-through-a-tube trick. The air you pull in through a tube is at ordinary air pressure; the water outside your body pushes in with a pressure equal to that of the air, plus the pressure of the water itself; a pressure that depends upon your depth below the surface.

This means that there is excess pressure pushing your chest inward and, in inhaling, you have to expand your chest cavity against that pressure. If you're deep enough, the water pressure becomes so high that you simply cannot expand your chest against it. In that case, you cannot inhale and you suffocate.

Yet why leave the air at its ordinary pressure? Suppose you must work at the bottom of a river, as when you are constructing the foundations of a bridge. One way of doing it would be to construct something like a bucket, which is put into the water, open end down. The water comes in but does not fill the bucket, of course. There is trapped air inside which is compressed by the incoming water until it exerts the same pressure downward that the water exerts upward.

What's more, you can push the water out of the bucket altogether if you pump more air into the bucket—air that is compressed to the point where it possesses the pressure of the water.

If you imagine a kind of inlet into the bucket through which workers can enter, you have a "pneumatic caisson."

Workers must be subjected by stages to more and more air pressure until they are at the pressure under which they will be working. The reason for the gradual increase in air pressure is to allow the pressure within the tissues of the body to equalize. Once the equalization has taken place, the caisson workers can move about freely, unaware of the additional pressure upon their bodies.

And yet there are difficulties. To see what those are, we have to consider what happens when air passes from our lungs into our body. The air doesn't do so as gaseous air; we can't use gases as gases.

What happens is that the air dissolves in the fluid layer coating the tiny vacuoles of the lungs and, in solution, diffuses across the thin lung membrane and the thin capillary membrane, passing, in this way, into the bloodstream. The dissolved gases are carried by the bloodstream to the tens of billions of cells in the body, and at each of those cells, the dissolved gases diffuse into the cells for possible use.

Air is not, however, a pure substance, and it does not dissolve as a single material. It consists of different gases, and each gas dissolves to a different extent.

The four chief gases of dry air, and their percentage by volume, are as follows:

gas	*percent by volume*
nitrogen	78.084
oxygen	20.946
argon	0.934
carbon dioxide	0.033

Various gases present in trace quantities make up the remaining 0.003 percent. In addition there are variable quantities of water vapor in the air as well as various kinds of dust particles. None of this is important in the following discussion and we will stick to the four major gases.

Suppose a pure sample of each of the four gases is thoroughly mixed with 100 milliliters (6.1 cubic inches) of pure water at 0° C. and 1 atmosphere pressure. How much of each gas will dissolve? Here is the answer in milliliters, which will measure the amount dissolved as percent by volume.

gas	*percent dissolved by volume*
nitrogen	2.33
oxygen	4.80
argon	5.60
carbon dioxide	171.3

This doesn't look so bad. Water can dissolve one-fiftieth of its own volume of nitrogen, about one-twentieth of its volume of either oxygen or argon, and nearly twice its volume of carbon dioxide.

It isn't entirely fair, however, to compare volumes of water with volumes of gas. Water is a liquid and quite dense, while nitrogen and the others are gases and very rarefied in comparison. While 100 milliliters of water weighs 100 grams, 100 milliliters of nitrogen weighs only 0.125 gram. Consequently, instead of measuring solubility as percent by volume, thus giving an unfairly impressive advantage to the gases, let us measure it as percent by weight.

How many grams of each gas will dissolve, under the conditions earlier expressed, in 100 grams of water? The answer is:

gas	*percent dissolved by weight*
nitrogen	0.0029
oxygen	0.0068
argon	0.0100
carbon dioxide	0.339

These are the results, remember, when a different batch of pure water is each mixed with a pure sample of a different one of these gases. Air is not, however, a pure gas, but a mixture of gases that all compete for a chance to dissolve. The larger the percentage of a particular gas in the air, the greater the chance of its dissolving, and the closer it approaches to its total solubility.

If a fresh batch of water is well mixed with a sample of air, then allowing for the various percentages of its components, 100 grams of water will contain:

gas	*percent dissolved out of air, by weight*
nitrogen	0.0023
oxygen	0.0014
argon	0.0001
carbon dioxide	0.0001

This certainly doesn't seem like much. Cold water, well mixed with air, ends up with 1 gram of oxygen for every 70,000 grams of water, and fish have to live on that small quantity of dissolved oxygen. They have to suck a lot of water past their gills to get enough oxygen out of it to fulfill their energy needs. But they manage.

As the temperature goes up, the solubility of all gases, including oxygen, goes *down.* At 25° C. (77° F.), 100 grams of water will hold only 0.00081 gram of oxygen, only five-ninths of what it holds at 0° C. That puts fish, and sea life generally, under an additional strain, and it is not surprising that the cold polar seas are considerably richer in sea life than the warm tropical seas.

Nor need we feel blessed because we live directly on the oxygen in the air, because in a way, we don't. The oxygen is no good to us until it dissolves in the thin film of water that lines the vacuoles of our lungs and that water is at a temperature of about 37° C. We are nowhere near as lucky as the polar fish, and we can thank our efficiency at quickly dissolving the oxygen and as quickly sucking it out into our red blood corpuscles for our ability to live.

Solubility of gases also varies directly with pressure. If air is compressed to five times its normal density and pressure, as would be true for people working in a caisson 41 meters below the water surface (or for people diving that far down in a scuba apparatus), then each gaseous component of the air would dissolve in five times the quantity it would in the normal pressure of sea-level air.

This can cause problems.

Since air is one-fifth oxygen, 5 atmospheres of pressure produces the equivalent of 1 atmosphere of oxygen. That much oxygen would, eventually, produce serious results due to oxygen toxicity, and this is something that must be watched for.

Again, there is a tendency at high pressures to resist breathing. One presumably gets a feeling of weariness at having to push that dense air in and out of the lungs especially when one gets the feeling one has enough oxygen with shallow breathing. This does stave off oxygen toxicity, but it results in the accumulation of carbon dioxide in the body, which produces carbon dioxide intoxication, with symptoms such as headache, dizziness, and worse. This, too, contributes to limiting the length of time people can work in caissons.

The matter of nitrogen or argon would seem to be benign. The body has no occasion to use either. Our body fluids hold both gases in solution at all times; they don't bother the body and the body doesn't bother them.

If, under compression, the body fluids hold, say, five times the normal quantity of dissolved nitrogen and argon, that still isn't much in an absolute sense and we might expect it wouldn't be bothersome. Yet it could be.

At quite high compressions, the dissolved nitrogen can produce nitrogen narcosis, which produces euphoria, overconfidence, and a decline in mental ability and in judgment. This is a very dangerous combination and it is, in fact, very similar in its effect to alcoholic intoxication. Divers who penetrate too far below the ocean surface are apt to experience what Jacques Cousteau calls "rapture of the deep," and the resulting carelessness may lead to a happy death.

Argon can produce similar results, but it is present in only one-twentieth the quantity of nitrogen and it can be ignored.

The most serious problem in compressed-air work or play arises when people who have been subjected to compressed air are brought back to normal pressures.

As the air pressure diminishes, the body fluids can hold less and less nitrogen in solution. The nitrogen must diffuse out of the cells through the cell membranes, into the bloodstream (which carries its own excess), and then out of the body by way of the lungs. This can be done without harm but it is a slow process.

Fortunately, the nitrogen is slow in coming out of solution in the first place. Even if the body is brought back to normal pressure so quickly that the excess nitrogen has not had time to bleed off, it continues to come out slowly and ooze out of the body safely. When people have not been exposed to too great a pressure of compressed air, decompression can take place at the rate that original compression took, with pressure in and out equalizing and no problem arising from the nitrogen excess.

As time went on, however, people worked at deeper and deeper levels under more and more air compression and built up higher and higher levels of nitrogen excess in their tissues.

When that happened it became possible for decompression to be too rapid. The nitrogen would come out of solution too rapidly for diffusion to get rid of it. The nitrogen would accumulate in tiny bubbles, which could do enormous harm.

If the bubbles formed in the joints and around nerves, they could produce agonizing pain. In the blood, they could suffocate. In the spinal cord, they could paralyze. In the brain, they could cause blindness or convulsions. If the condition were bad enough it could do permanent damage, or kill.

The symptoms could set in anywhere from one to eighteen hours after decompression and the condition is called *decompression sickness, caisson disease,* or *the bends.*

When caisson work first plunged to significantly great depths, decompression sickness became the scourge of underwater construction. The first large steel bridge was the Eads Bridge built from 1867 to 1874 over the Mississippi River at St. Louis. It was named for its builder, James Buchanan Eads (1820–87). To sink the foundations, excavations had to be made to the then unprecedented depth of 30 meters. There were about 600 men engaged and 119 had bad cases of decompression sickness. Fourteen of them died.

Then, between 1869 and 1883, the Brooklyn Bridge, connecting Manhattan and Brooklyn, was built. It was the first of the great suspension bridges and almost everything about it was experimental.

The man in charge of the construction was Washington Augustus Roebling (1837–1926), the son of the designer, who was killed in an accident at the very beginning of the construction.

Despite precautions, more than a hundred cases of decompression sickness resulted, and one of the casualties was Roebling himself. Roebling, a totally dedicated person, insisted on supervising every facet of the project whatever the dangers involved. At one point he remained for twelve consecutive hours in a caisson, and when he finally lapsed into unconsciousness he was brought to the surface—too quickly.

Decompression sickness permanently damaged his body. He was confined to his apartment, which overlooked the site where the bridge was being built. From there, he watched and supervised the construction, with his wife serving as his mobile unit, carrying orders to the engineers and foremen and bringing back reports. He lived to see the Brooklyn Bridge complete and for forty-three years more, but never recovered his health completely.

The proper way of treating or preventing decompression sickness was first suggested in 1907 by John Scott Haldane (1860–1936).

He suggested that people decompress in stages. It is usually safe to reduce the pressure to no less than one-half its previous value at some reasonable speed, but then people must stay at that pressure until the nitrogen in the body fluids has reached a new and lower equilibrium. Then the pressure is reduced another stage and there is another wait and so on. It is a tedious method, but anyone who has witnessed what decompression sickness can do to a person would rather go through the tedium any number of times than have decompression sickness once.

If, through carelessness or accident, decompression has been too fast and the symptoms of decompression sickness begin to appear, a person must be placed under compressed air at once, to dissolve the gas bubbles once again, and then decompression by stages can take place.

One possible way of avoiding the ill effects of compressed air is to replace the nitrogen with some other gas that would be even more inert than nitrogen and therefore less dangerous in various ways. The obvious choice was helium.

Helium is the most inert of all substances. It is so inert that it resists even solution, so that it is the least soluble of all known gases. Thus, 100 milliliters of water will dissolve 0.94 milliliters of helium, only two-fifths of the volume of nitrogen it will dissolve.

This looks hopeful since it means that if bubbles formed during decompression, the volume of helium bubbles formed would be only two-fifths that of the nitrogen bubbles formed under similar conditions and that should produce correspondingly less damage.

Unfortunately, that is not all there is to the story. Helium's greater inertness and its smaller atoms make it readier than nitrogen to leave the solution. This means that though there is less bubbling in the case of helium, the bubbles form faster, and the two tendencies largely cancel each other. Helium does not, therefore, liberate people from the fear of decompression sickness. In fact, decompression in the case of helium-oxygen mixtures must take place by smaller stages than in the case of nitrogen-oxygen mixtures. (I suspect, however, that helium reaches equilibrium faster so that the stay at each stage is shorter, and that is a good thing.)

Helium has advantages that have nothing to do with decompression sickness. Helium has only one-ninth the intoxicating effects of nitrogen so that one can go deeper without danger of suicidal "rapture." Then, too, helium is one-seventh as dense as nitrogen. Air in which helium replaces nitrogen is only a third as dense as ordinary air. This reduces the viscosity of compressed air and makes it less difficult to push it in and out of the lungs. People breathe more readily and the chance of carbon dioxide intoxication is lowered.

For these reasons the use of helium does make it safe to go to depths beyond those where one can go with ordinary compressed air. Helium is used for scuba divers who intend to spend considerable time below 45 meters.

Whereas divers breathing ordinary compressed air dare not sink below 90 meters for even brief intervals, helium makes it possible to go down to 150 meters for considerable periods and to still greater depths under special conditions.

Yet all these divings represent only the skin of the ocean. Even a depth of 150 meters is less than one-seventieth the greatest depth of the ocean.

What is the pressure at the great depths?

That depends upon temperature, salinity, currents, compressibility, but the heck with that. We can get reasonably close if we merely start with 1 atmosphere for the air itself and then add another atmosphere for every 10.332 meters.

The greatest depth of the ocean is to be found in the Marianas Trench in the Pacific, where a depth of 11,033 meters (about 6.9 miles) has been reported. There the water pressure would be something like 1,070 atmospheres.

To see what that means in a familiar unit, water pressure at the bottom of the deepest trench is equal to 7.85 *tons* per square inch.

It would seem, if we didn't know better, that life would be impossible under such conditions, but, of course, that's not so. Life has been found at all depths in the ocean, even the lowest. Internal and external pressures balance and that's what counts.

We won't leave the subject, though, without giving the water pressure in the deepest trench in the metric system. The maximum water pressure in the ocean is:

	1,084,000,000	dynes per square centimeter
or	1,084	bars
or	1,084,000	millibars
or	108,400,000	newtons per square meter
or	108,400,000	pascals

And this isn't the largest pressure you can find under natural circumstances on earth—but that is something I will discuss in the next chapter.

3 * UNDER PRESSURE

As some of you may have noted with varying degrees of resignation or indignation, I am a devotee of wordplay. I get a great deal of innocent fun out of such things, for the groans of the audience are music to my ears—especially since my reputation is so vile in this respect that a pun is suspected even when I had intended none.

Recently I said, in the course of a discussion on population growth, that it took a woman nine months to have a baby and that no labor-saving device had yet been invented that—

And at this point, a chorus of groans rose in the air and I had to think back on what I had said to find the wit. Having found it, I smiled modestly as though I had intended it all along.

It changes from fun to a serious duel, however, when one punster finds another. The other does not enjoy your cleverness. No, he sits there under pressure and tries to top you with another pun that is related to yours. Then, when the shock of his riposte strikes you, you must churn your brains wildly to come back with something else with a minimum of delay and so the duel escalates itself, until the two of you have to be led away by kindly bystanders and put to bed with ice bags on each fevered brow.

I name no names, which is why I am careful not to mention Mark Chartrand, director of the Hayden Planetarium and, later, of the National Space Institute.

Recently, I met a pleasant dentist and his attractive wife and they and Janet and I had several interesting conversations. On one occasion, after discussing science fiction movies, with special attention to *Star Wars,* he said, as he departed, "The dental blessing on you."

"Oh?" said I. "What's that?"

And he said, "May the floss be with you!"

I laughed so hard I didn't get a chance to top him, which filled me with chagrin. Had I had my breath available to me, I could instantly have come back with the gambler's blessing, "May the horse be with you!" or the henpecked husband's blessing, "May divorce be with you!" or the sinner's blessing, "May remorse be with you!" or even the blabbermouth's blessing, "May discourse be with you!" and so on.

Oh, well, I was spared the duel and the pressure and that means enough pressure is left me to carry on the discussion begun in the preceding two chapters.

From the earliest days of chemistry it was quite understood that varying temperature affected the properties of substances and the manner in which they underwent chemical reactions. Raise the temperature of ice and it melts; raise the temperature of a mixture of hydrogen and oxygen and it explodes.

Changes in pressure will also alter properties and reactions, but it is a great deal more difficult to change pressure significantly than temperature. Prehistoric human beings could change temperature merely by use of fire, and had been cooking food, smelting ores, baking clay, making glass—all as a result of chemical changes produced or accelerated by rising temperatures.

Yet they had no way of altering pressure substantially except by hammering. The first to study changes in property with pressure systematically was the British scientist Robert Boyle (1627–91). In 1662, he experimented with air under the pressure produced by the weight of mercury, and found that air was compressible and that its volume shrank in proportion as pressure increased.

This was a strong piece of evidence in favor of the fact that gases were composed of tiny particles of matter sparsely strewn through vacuum. Pressure crowded the particles closer together, reducing the overall volume, but when the pressure was removed the particles separated, as though by a natural springlike reaction, to their original volume. (Boyle called it "the spring of the air.")

Eventually, when the atomic theory was accepted, this was precisely the explanation accepted to explain the compressibility of air, and of gases generally.

Liquids and solids could not be measurably compressed in this manner by the pressures available to Boyle and those who immediately followed him, and this made sense, too. The ultimate particles of matter (that is, the atoms, and the atom combinations called molecules) were in contact in liquids and solids, and once they were in contact it might be supposed that they could not be moved closer together and there could be no further compression.

In 1762, however, the English physicist John Canton (1718–72) demonstrated before the Royal Society that water was indeed slightly compressible. After all, even if water consists of atoms and molecules in contact, there are different types of packing, some more compact than others, and atoms and molecules might be distorted or even outrightly compressed into closer fits.

Unfortunately, once the atomic theory was accepted, it was also fashionable to think of atoms as truly ultimate—indivisible, undistortable, impenetrable. There was no experimental evidence for this, but the ancient Greek philosophers who had first dreamed up atoms had asserted this, and nineteenth-century chemists found it difficult to abandon the pied pipers of Hellenism.

The result was that it was casually accepted that liquids and solids, with atoms and molecules in contact, were incompressible despite experimental results to the contrary. (I was taught as much when I was young by high school teachers who knew no better, and discovering that this was not so came as a shock to me at first.)

It was true, however, that although liquids and solids were compressible, they were not *easily* compressible, and throughout the nineteenth century, pressure experiments continued to deal with gases almost exclusively. Pressure did not merely compress gases. Some gases (not all) were liquefied by pressure alone, or at most, by pressure and moderate cooling. It was the effort to extend this effect to all gases that first led scientists into the realm of high pressure.

The French physicist Emile Hilaire Amagat (1841–1915) was a pioneer in this respect. By applying mechanical pressure to a small volume and devising seals that were particularly efficient, he managed to reach pressures as high as 3,000 atmospheres in the 1880s.

But even the best seals leak under sufficient pressure. Attempts to reach more than 3,000 atmospheres failed because the seals gave way—and yet it was no mean accomplishment. Amagat reached pressures three times those in the deepest portion of the oceanic abyss.

In 1905, the American physicist Percy Williams Bridgman (1882–1961) was working for his Ph.D. at Harvard. He was studying the behavior of certain optical phenomena under the influence of pressure and began to interest himself in the problem of reaching higher and higher levels of pressure.

He worked out ingenious seals that would retain fluid under more and more stringent conditions. He soon reached a pressure of 12,000 atmospheres and then, in successively improved devices, went to 20,000, then 30,000, then 50,000, then 100,000, and then, finally, an occasional 425,000 atmospheres. In 1946, he earned a Nobel Prize in physics for his high-pressure work.

Bridgman was the first who could fiddle around usefully with the atomic arrangements of liquids and solids, and this opened up the possibility of some pretty dramatic accomplishments.

For instance, graphite and diamond are both made up of carbon atoms only. Diamond, however, has a density that is 1.56 times as great as that of graphite. That is because the carbon atoms in diamond are packed more compactly than those in graphite.

Suppose graphite is placed under pressure. The carbon atoms, pushed together with greater and greater force, finally yield and take up the more compact diamond configuration. It takes enormous pressure to do this, however, so it is no wonder that attempts prior to Bridgman's time to turn graphite to diamond all failed.

Even Bridgman found the pressures he disposed of incapable of doing the job unaided. If, however, the graphite is heated, the grip of each carbon atom on its neighbors is weakened, and the pressure necessary to force those atoms to rearrange into the diamond configuration is reduced to manageable proportions. In 1955, a combination of high temperature and high pressure allowed the formation of diamonds from graphite for the first time.

Once the pressure is removed, the tendency is for the carbon atoms in diamond to revert to the graphite configuration in a kind of "spring of carbon" effect. However, the carbon atoms are held together so tightly once they're in the diamond configuration that they can revert to graphite only with excessive slowness.

Diamonds are unstable, in the strictest sense of the word, but it is a very slow-motion instability that makes them as good as stable for all practical purposes. Something that is unstable in theory and stable in practice is said to be *metastable.*

The formation of diamond was dramatic indeed, but earlier, Bridgman had done work on ordinary water that, in its way, produced the more startling results.

Water freezes into ice at 0° C. (32° F.). Ice, however, has a density that is only 0.92 that of liquid water. The molecules of water (each made up of two hydrogen atoms and an oxygen atom) are loosely arranged in ice, though they are, after a fashion, in contact; and more compactly arranged in water.

This means that just as pressure alone can change graphite into the more compact diamond, so pressure alone can change ice into the more compact water. What's more, the bonds holding the water molecules to each other in ice are far feebler than those holding the carbon atoms together in graphite. Pressure therefore effects the ice-to-water change much more easily and readily than it would effect the graphite-to-diamond change.

Even a relatively small amount of pressure applied to ice that is not too far from the freezing point will cause it to melt, and it will have to be cooled down to a lower-than-normal temperature to make it freeze again. It is calculated that each additional atmosphere of pressure over and above the normal reduces the freezing point by 0.0075° C. (0.0135° F.). If you place ice under a pressure of 135 atmospheres, it will freeze at −1° C. (30.2° F.).

The sharp edge of the blade of an ice skate supports the total weight of an ice skater. Since the blade covers only a tiny area of the ice, the weight of the skater, concentrated on that area, produces a huge, though local, pressure.

If the weather isn't too cold, the freezing point under the blade is lowered to below the atmospheric temperature and the ice melts. There is thus a thin film of water under the blade which acts as a lubricant and makes ice skating a good deal smoother than it would be if this phenomenon didn't exist.

The act of skating does not, however, melt the ice on the pond permanently, for when the skate leaves the area it covered, the water film it formed returns to ordinary pressures and freezes again at once. Water is present only immediately under the skate blade and nowhere else.

Again, suppose a weight were suspended from a loop of thin, strong wire which was thrown over a block of ice maintained at a temperature just below its freezing point.

Under the thin wire, the pressure is enormous and the ice melts. The wire sinks through the water, which freezes above the wire. The wire continues to melt the ice underneath and to sink, and finally works its way entirely through the block of ice, which remains as solid and unbroken as ever.

Suppose you placed very large pressure on the ice and lowered the temperature steadily to prevent it from melting under that pressure. At a pressure of 2,047 atmospheres, the freezing point of water is —22° C. (—7.6° F.)—and then an odd thing happens. The water molecules in the ice rearrange into a more compact solid form and you no longer have ice; at least not our ordinary ice. It is customary to call ordinary ice ice I, while the new form of ice is ice III.

Ice III has a more compact molecular arrangement than ice I and even than liquid water. Ice III is denser than water and will sink in it, whereas ice I (as we know) floats.

If the pressure is increased still further, the tendency now is for water to be compressed into the more compact ice III so the freezing point rises. At 3,417 atmospheres, the freezing point of water is up to —17° C. (1.4° F.) and ice III then squeezes into the still more compact and dense ice V. (You can tell these shifts, by the way, by the sudden changes in volume that take place.)

The freezing point rises still more sharply thereafter with pressure increase and at 6,175 atmospheres, it is 0.16° C. (32.29° F.) or just about the ordinary freezing temperature we know and love—except that the pressure is enormous and ice V crushes into a slightly more compact ice VI.

At the enormous pressure of 21,700 atmospheres, the freezing point is at 81.6° C. (178.9° F.) and ice VII forms. Its density is 1.7 times that of water (and 1.85 times that of ice I). Its freezing point rises yet more rapidly with increasing pressure, and at somewhere around 23,000 atmospheres, ice VII's melting point is at 100° C. (212° F.), which is the boiling point of water under ordinary conditions. Imagine solid water at the temperature of steam.

Haven't I omitted ice II and ice IV in all of this?

Well, ice II forms at pressures that suffice to form ice III, but at lower temperatures. Ice II never melts into water. If you imagine it formed at low temperatures and then slowly raise the temperature without altering the pressure, ice II will reach a point where it will change into ice III or ice V (depending on what the pressure is) and it is these latter that eventually melt into water as the temperature is further raised.

As for ice IV, it is only metastable. It forms at low temperatures but will change spontaneously into other ice forms in time.

There is also an ice VIII that forms at low temperatures at very high pressures. If its temperature is raised, it turns into ice VI or ice VII and it is these that eventually melt to water with a further temperature rise.

Of all the eight ice forms (sometimes referred to as *ice-o-morphs*), only ice I is less dense than water, only ice I is stable at ordinary pressures and only ice I exists in nature. The ocean pressures even at their deepest are only half what is required to form ice III, the next more compact form. And if the ocean's depth were doubled and sea-floor pressure became sufficient, the temperature wouldn't be low enough for the task.

What is more, all those compact ice forms, if the pressure were released, would quickly revert to ice I. The weak bonds between the molecules would not suffice to hold them compactly against the tendency to relax and spread out under low pressure. Ice is not diamond.

(In one of his novels, Kurt Vonnegut invented an "ice IX," which was metastable, retained its integrity at ordinary temperatures and pressures, and which, what's more, would serve as a seed that would convert earth's entire water supply into ice IX and, therefore, make all life impossible. You will be glad to know that however interesting it made the novel, ice IX or any ice form with that set of properties cannot exist in the real universe.)

But wait. If we are talking about naturally occurring high pressure on earth, the ocean is by no means the limit. The ocean is made up of water which has a depth of only 11 kilometers (7 miles) at most.

The solid ball of the earth, on the other hand, goes 6,378 kilometers (3,963 miles) straight down to the center, calculating from a point on the equator. The pressure rises steadily as we imagine ourselves delving underground. The average density of the surface rocks is 2.8 times that of water, and if the density of the earth's structure remained that way all the way down, then 1 atmosphere would be added for each 3.7 meters we would go downward.

At a depth of a little over half a kilometer, the pressure would be high enough to form ice III, but ice III requires a temperature no higher than —22° C., and at a depth of half a kilometer, the temperature of the rocks is about 44° C. (112° F.), which is considerably too high.

At a depth of 6.2 kilometers, the pressure is high enough to form ice VII under conditions where it is solid at the ordinary boiling point of water, but down there, the temperature of the rocks is well above that boiling point and stands at about 200° C. (390° F.).

Even if there were water low enough in the rock of the solid earth to experience pressures to form compact ice forms, the temperatures would always be too high. Again, then, I say that only ice I occurs anywhere on earth.

If the density of the earth were a uniform 2.8 times that of water all the way through, then the density at the center of the earth would be something like 1,700,000 atmospheres, or about 1,600 times that at the bottommost trench of the ocean.

Actually, though, the increasing pressures force the atoms and molecules in the rock to move closer together and to take up more compact configurations. By the time that a depth of 2,900 kilometers (1,800 miles) is reached, the density of the rock has more than doubled over the surface value and is about 5.8 times that of water. (These changes in density can be calculated from the speed with which earthquake waves travel through the earth at different depths—something we don't have to go into here.)

At the depth of 2,900 kilometers (1,800 miles) there is a sudden sharp increase in density to about 9.5 times that of water and it then increases smoothly until it is about 11.5 times that of water at the center of the earth. The portion of the earth down to that 2,900-kilometer rise is the *mantle,* while the portion of the earth below that rise is the *core.*

Taking into account the increasing density with depth, the weight of overlying rock at any given point (and therefore the pressure) must be higher than it would be if the rock were of uniform density. The figure of 1,700,000 atmospheres at the center is too low, therefore.

Actually, the central pressure is about double this and can be put at 3,500,000 atmospheres. This amounts to 25,718 *tons* per square inch, or 36,000,000,000 kilograms per square meter. If we use the SI unit system then we can say the pressure at the earth's center is 355,000,000,000 newtons per square meter, or that many pascals.

That sudden increase at a depth of 2,900 kilometers (1,800 miles) can only take place if there is a sharp alteration in earth's composition. The rock must come to an end and something denser must take its place. The only substance that can be substantially denser would have to be largely or entirely metal, and the most common metal in the universe that is dense enough to fit the data is iron.

As it happens, a number of meteors are almost pure iron and nickel in a ten-to-one ratio (nickel is a metal very similar to iron in chemical properties). If such meteors are remnants of an exploded planet (a popular and dramatic theory in past decades), then this may be a good sign that the dense central core of the earth is a nickel-iron alloy. This was, in fact, first suggested as long ago as 1866 by the French physicist Gabriel Auguste Daubrée (1814–96).

What's more, the core is hot enough for the nickel-iron to be liquid, though the mantle is solid. This is not surprising, for the mantle is at a somewhat lower temperature than the core and rock has a higher melting point than iron.

But if the earth consists of two portions, a rocky outside wrapped about an iron inside, this must have developed. It can't be that first a mass of nickel-iron accumulated and then, when that was done, a mass of rock formed about it. No, bits and pieces of all types must have come together into a more or less chaotically distributed mélange of different substances, whereafter the iron must have separated out and settled to the bottom.

Of course, it's easy to say "must have," but did the iron actually do so? It's difficult to come to any sensible decisions when all the evidence we have at our disposal consists of earthquake waves and shaky reasoning based on meteorites and the presumed behavior of materials at high temperatures and pressures.

We need some more direct evidence, and Peter M. Bell of Carnegie Institution is trying to supply it. He has made use of a device which squeezes materials between two diamonds (diamond being the hardest substance known) and with it has managed to reach pressures of 1,500,000 atmospheres, over two-fifths that at the earth's center. He believes it is possible for the instrument to go to 17,000,000 atmospheres before the diamonds themselves fail.*

Furthermore, laser beams can be focused through the transparent diamonds to raise the temperature of the material being compressed. Thus, Bell can study the behavior of substances under the kind of temperatures and pressures characteristic of the environment deep within the body of the earth.

Bell's pressures have succeeded in compressing various minerals characteristic of earth's crust into very compact structures of the kind found naturally in a mineral called *perovskite* (a form of calcium titanate). This perovskite structure, he believes, is characteristic of the mantle at depths of more than 800 kilometers (500 miles).

While most types of atoms take up their place in the perovskite structure, iron atoms do not. Bell found that when iron-containing minerals are placed under high temperature and pressure in his apparatus, the iron atoms leak out of the perovskite structure and collect as the pure metal.

One can imagine, therefore, earth forming under the collision of rocky fragments, and growing larger. As it grows larger, the pressures and temperatures within grow higher. As they do, more and more of the compact perovskite structure forms and any iron atoms present are squeezed out and, eventually, because of their high density, these collect at the center of the earth.

If this is so, then I wonder how the nickel-iron meteorites formed? If they are not part of the core of an exploded planet (and modern astronomers don't seem to want to accept exploded planets as part of the history of the solar system), how else could that iron have collected?

* At the California Institute of Technology, shock waves are used to produce temporary pressures that are higher still—up to 75,000,000 atmospheres, I believe.

Then, too, even Bell's data are a little on the indirect side. It would be delightful to obtain a sample of the core and actually analyze it. The problem is, though, how can one get down to the core to get our sample?

Actually, there is a way. The core can be subjected to analysis in place. The key to this is the neutrinos, which are constantly emerging from the sun and which (if aimed properly) pass completely through the earth without trouble.

When the sun is above the horizon, the neutrinos reach us and our instruments after having passed across 150 million kilometers (93 million miles) of relatively empty space. When the sun is below the horizon, the neutrinos reach us and our instruments after having passed through that distance of space, *plus* several thousand kilometers of earth's structure. Exactly how much of earth's structure has been traversed and through what depths the neutrinos have passed before reaching us depends on how far below the horizon the sun is.

Most of the neutrinos passing through the earth do not interact with any of the matter they penetrate, but a few would, and the details of interaction might vary with temperature, pressure, and composition.

If we can develop instruments that will detect neutrinos and obtain a detailed energy spectrum of them, then by comparing daytime neutrinos and nighttime neutrinos, we could conceivably achieve a complete analysis of the earth's structure at every level.

It's a nice thought—but easier to talk about than to do—and I believe Hal Clement made use of the idea in one of his novels that first appeared in 1970.

4 ★ THE WORD I INVENTED

Robotics has become a sufficiently well developed technology to warrant articles and books on its history and I have watched this in amazement, and in some disbelief, because I invented it.

No, not the technology; the word.

In October 1941, I wrote a robot story entitled "Runaround," first published in the March 1942 issue of *Astounding Science Fiction,* in which I recited, for the first time, my Three Laws of Robotics. Here they are:

1. A robot must not injure a human being or, through inaction, allow a human being to come to harm.
2. A robot must obey the orders given it by human beings except where those orders would conflict with the First Law.
3. A robot must protect its own existence, except where such protection would conflict with the First or Second Law.

These laws have been quoted many times by me in stories and essays, but what is much more surprising is that they have been quoted innumerable times by others (in all seriousness) as something that will surely be incorporated in robots when they become complex enough to require it.

As a result, in almost any history of the development of robotics, there is some mention of me and of the Three Laws.

It is a queer feeling to know that I have made myself into a footnote in the history of science and technology for having invented the foundation of a science that didn't exist at the time—and that I did it at the age of twenty-one.

The Three Laws, and the numerous stories I have written that have dealt with robots, have given many people—from enthusiastic teenage readers to sophisticated editors of learned magazines in the field—the idea that I am an expert on robots and computers. As a result, I am endlessly being asked questions about robotics.

What I will do, then, is write a question-and-answer essay on the subject. It will take care of just about all the major questions I am forever being asked and it should make it unnecessary for anyone to have to ask me any questions on the subject again.*

1. Dr. Asimov, how did you come to be such an expert in the field of robotics?

Alas, I am not an expert, and I never have been. I don't know how robots work in any but the vaguest way. For that matter, I don't know how a computer works in any but the vaguest way, either. I have never worked with either robots or computers, and I don't know any details about how robots or computers are currently being used in industry.

I don't take pride in this. I merely present it as a fact. I would like

* But I am dreaming. The questions will continue, I know.

to know all about robots and computers but I can only squeeze so much into my head, and though I work at it day and night with remorseless assiduity, I still only manage to get a small fraction of the total sum of human knowledge into my brain.

2. In that case, Dr. Asimov, how did you come to write so many robot stories, considering that you know nothing about the subject?

It never occurred to me that I had to. When I was reading science fiction in the 1930s, I came across a number of robot stories and learned what I had to know on the subject from them.

I found out that I didn't like stories in which robots were menaces or villains because those stories were technophobic and I was technophilic. I did like stories in which the robots were presented sympathetically, as in Lester del Rey's "Helen O'Loy" or Eando Binder's "I, Robot."

What's more, I didn't think a robot should be sympathetic just because it happened to be nice. It should be engineered to meet certain safety standards as any other machine should in any right-thinking technological society. I therefore began to write stories about robots that were not only sympathetic, but were sympathetic *because they couldn't help it*. That was my contribution to this particular subgenre of the field.

3. Does that mean you had the Three Laws of Robotics in mind when you began writing your robot stories?

Only in a way. The concept was in my mind but I wasn't smart enough to put it into the proper words.

The first robot story I wrote was "Robbie" in May 1939, when I was nineteen. (It appeared in the September 1940 *Super-Science Stories*, under the title of "Strange Playfellow.") In it, I had one of my characters say, about the robot hero, "He just can't help being faithful and loving and kind. He's a machine—*made so.*" That was my first hint of the First Law.

In "Reason," my second robot story (April 1941, *Astounding*), I had a character say, "Those robots are guaranteed to be subordinate." That was a hint of the Second Law.

In "Liar," my third robot story (May 1941, *Astounding*), I gave a version of the First and Second Laws, when I said the "fundamental law" of robots was: "On no conditions is a human being to be injured in any way, even when such injury is directly ordered by another human."

It wasn't, however, till "Runaround," my fourth robot story, that it all came together in the Three Laws in their present wording, and that was because John Campbell, the late great editor of *Astounding,* quoted them to me. It always seemed to me that John invented those laws, but whenever I accused him of that, he always said that they were in my stories and I just hadn't bothered to isolate them. Perhaps he was right.

4. But you say you invented the term robotics. *Is that right?*

Yes. John Campbell, as best I can remember, did not use the word in connection with the Three Laws. I did, however, in "Runaround," and I believe that was its first appearance in print.

I did not know at the time that it was an invented term. The science of physics routinely uses the *-ics* suffix for various branches, as in mechanics, dynamics, electrostatics, hydraulics, and so on. I took it for granted that the study of robots was robotics.

It wasn't until a dozen years later, at least, that I became aware that *robotics* was not listed in the second edition of Webster's New International Dictionary or (when I quickly checked) in any of the other dictionaries I consulted. What's more, when Webster's third edition was published, I looked up *robotics* at once and still didn't find it.

I therefore began saying that I had invented the word, for it did indeed seem to me that I had done so.

In 1973, there appeared *The Barnhart Dictionary of New English Since 1963,* published by Harper & Row. It includes the word *robotics* and quotes a passage from an essay of mine in which I claim to have invented it. That's still just me saying so, but at least the lexicographers didn't cite earlier uses by someone else.

The word is now well established and it is even used in the titles of magazines that are devoted to the technology of robots. To be candid, I must admit that it pleases me to have invented a word that has entered the scientific vocabulary.*

5. I frequently hear your robots referred to as positronic robots. *Why positronic?*

When I first began writing science fiction stories, the positron had been discovered only six years before as a particle with all the properties of an electron except for an opposite charge. It was the first (and, at that time, still the only) bit of antimatter that had been discovered, and it carried a kind of science fictional flavor about it.

* *Psychohistory,* which I also invented, has entered the scientific vocabulary, but, alas, not in the sense of my invention.

That meant that if I spoke of *positronic robots* rather than *electronic robots,* I would have something exotic and futuristic instead of something conventional.

What's more, positrons are very evanescent particles, at least in our world. They don't survive more than a millionth of a second or so before they bump into one of the electrons with which our world is crowded, and then the two annihilate each other.

I had a vision, therefore, of "positronic pathways" along which positrons briefly flashed and disappeared. These pathways were analogous to the neurons of the animal nervous system and the positrons themselves were analogous to the nerve impulse. The exact nature of the pathways was controlled by positronic potentials, and where certain potentials were set prohibitively high, certain thoughts or deeds became virtually impossible. It was the balance of such potentials which resulted in the Three Laws.

Of course, it takes a great deal of energy, on the subatomic scale, to produce a positron; and that positron, when it encounters an electron and is annihilated, produces a great deal of energy on the subatomic scale. Where does that positron-producing energy come from and where does the positron-annihilation energy go to?

The answer to that is that I didn't know and didn't care. I never referred to the matter. The assumption (which I didn't bother to state) was that future technology would handle it and that the process would be so familiar that nobody would wonder about it or comment upon it—any more than a contemporary person would worry about what happens in a generating plant when a switch is flicked and a bathroom light goes on.

6. You also talk about platinum-iridium brains. Why those metals?

Partly because that was a symbol of the value of the brains. Platinum and iridium are rarer than gold and, at the time I wrote the early stories, they were more valuable than gold, too.

In addition, platinum-iridium is about as inert as metal can possibly be, and I needed something inert as background for the flashing positronic pathways. The framework of the brain would have to be long lasting and static.

7. Talking about the positronic energies reminds me, Dr. Asimov, to wonder where your robots got the energy to do their work. Where?

I assumed some form of nuclear power (or "atomic power," as we called it in the 1930s).

When I wrote my first robot story in 1939, uranium fission was just being discovered, but, of course, I had not heard of it yet. That, however, didn't matter. From about 1900 on, it was perfectly well understood that there was a source of huge and concentrated energy in the interior of the atom. It was standard "believe-it-or-not" fare to be told that if all the energy in an ounce of matter could be extracted it would suffice to drive a large ocean liner across the Atlantic.

Consequently, the general science fictional thought was that some very small object, serving as an "atomic energy device," could be inserted into a robot and that would keep it running for millions of years, if necessary.

As the years passed and we learned a great deal about the practical aspects of nuclear energy, I might have yielded to the headlines of the moment and spoken wisely of uranium fission and cadmium rods and so on, but I did not do so. I think I was right in this. I maintained silence of the details of the energy source because it had nothing to do with the point of the stories, and that caused no reader discomfort that I am aware of.

8. In your earliest robot stories you made no mention of computers, yet surely the positronic brain is actually a very complex, compact, and versatile computer. Why did you not say so?

Because it never occurred to me to say so. As a science fiction writer, I was a creation of the science fiction of the 1930s, which was written by writers who built on what had gone before.

As it happened, the world of fiction had been full of humaniform objects brought to artificial life, including the Golem and Frankenstein's monster. There were also various "automatons" in human shape. Such things were in the air. Čapek invented the word *robot* for them, but the word was applied to a concept that had long existed.

Computers, on the other hand, were not really in the air until the first electronic computer was built during World War II. Earlier mechanical calculating devices were so simple that they gave absolutely no birth to the thought of "thinking machine."

Since my robot stories began just before World War II, computers were not part of my consciousness, and I did not either talk or think of them. Yet even so I could not help introduce computers, even though I did not know what I was doing.

In my very first robot story, my little-girl heroine encounters a "talking robot" which "sprawled its unwieldy mass of wires and coils through twenty-five square yards." When it spoke, there was "an oily whir of gears." I hadn't managed to work out the notion of electronics in its connection, but what I had was a kind of mechanical computer.

By the time I wrote my story "Escape" in November 1944 (it appeared in the August 1945 *Astounding* as "Paradoxical Escape"), I had another huge nonportable structure which I referred to as a "thinking machine" and called "the Brain." That was written before the first electronic computer, Eniac, came into existence.

Eventually, I did begin to write computer stories. I think the first of those was "Franchise," which appeared in the August 1955 *If*. Even then I never completely differentiated robots and computers and I feel I was right not to do so. To me, a robot was a mobile computer, and a computer an immobile robot. From here on in then, when I speak of "robots" in this chapter, please remember that I use it to include computers as well.

9. Come to think of it, why are robots humaniform? Surely that is not the most efficient shape.

Again, it's a matter of history. The robot is in the tradition of the "artificial man," which goes far back in the human imagination.

It is a matter of drama. What can be so supreme an achievement as to create an artificial human being, so that we have the mythical Greek inventor, Daedalus, constructing a brazen man, Talos, who served to guard the shores of Crete? Again, what can be so supreme a blasphemy as to attempt to mimic the Creator by devising an artificial human being, so that we have the hubris-and-atē of Victor Frankenstein?

With such a background, science fiction writers were unable to think of intelligent machines without making them humaniform. Intelligence and the human shape seemed too intimately connected to be separated. It was only with the rise of the electronic computer, which presents a kind of artificial intelligence without the involvement of any fixed shape, that robots were seen as mobile computers and no longer had to be humaniform.

Thus, the very successful R2D2 in *Star Wars* was shaped like a fire hydrant, and seemed very cute in consequence, especially, for some reason,* to the female portion of the audience.

And if we step into the world of real robots, the kind that are

* It has been pointed out to me that R2D2 had a phallic appearance.

being used in industry now, we have only the vaguest sign of humaniformity, if any at all. But then they are as yet very primitive and limited in the tasks they can perform. It is possible that as robots become more versatile and generalized in their abilities, they will become more humaniform.

My reasons for thinking so are two:

a. Our technology is built around the human shape. Our tools, our appliances, our furniture, are built to be used by human beings. They fit our hands, our buttocks, our feet, our reach, the way we bend. If we make use of robots with proportions like ours, with appendages like ours, which bend as we do, they can make use of all our tools and equipment. They can live in our world; they will be technologically compatible with us.

b. The more they look like us, the more acceptable they will be. It may be that one of the reasons that computers arouse such distaste and fear in many people is that they are nonhumaniform and are therefore seen as a dehumanizing influence.

10. Well, then, when do you think we will have robots like those you describe, as intelligent and versatile, and subject to the Three Laws?

How can one tell? At the rate that computer technology is now advancing, it doesn't seem to me to be impossible that within a century, enough capacity and versatility can be packed into a device the size of a human brain to produce a reasonably intelligent robot.

On the other hand, technological capacity alone may well prove insufficient. Civilization may not endure long enough to allow robots to reach such a stage. Or even if it does, it may turn out that the social and psychological pressures against robots will prevent their development. Perhaps my feeling that humaniform robots will seem friendly will prove wrong. They may prove terrifying instead (something which I take for granted in my robot stories, by the way).

Then, too, even if the technological capacity is there and if social resistance is absent, it may be that the direction of technology will be different from that which I originally imagined.

For instance, why should each robot have an independent brain with all the expense and risk of damage that would entail?

Surely, it would make more sense to have some central computer be responsible for the actions of many robots.* The central com-

* Such a possibility is mentioned in my most recent robot story, "The Bicentennial Man" (*Stellar Two*, 1976).

puter in charge of a squadron of robots could be any size since it would not have to be portable, and, while expensive, it would certainly not be as expensive as a squadron of separate and very compact brains. Furthermore, the central immobile computer could be well protected and would not run the risk of the kind of damage that would always be possible in the case of mobile robot brains.

Each mobile robot would, we might imagine, have a characteristic wavelength to which it would respond and through which it would be connected to its own portion of the central brain. Without a brain of its own it could be risked in dangerous enterprises much more readily. The disadvantage would be that it would depend on electromagnetic communication that could be interfered with, perhaps, by both natural and technological means. In other words, a malfunctioning or nonfunctioning robot would then be much more likely.

11. Since you mention the possibility of a malfunctioning robot, how safe are the Three Laws anyway? They seem to be ambiguous. How do you define a human being? What do you mean by harm?

The Three Laws are deliberately ambiguous. From the very first time I used them in "Runaround," it was the ambiguity that supplied me with a plot. I considered the definition of *harm* as early as my story "Liar!" and in my novel *The Naked Sun* (Doubleday, 1957) I even dealt with robotic murder, despite the Three Laws.

As to how a human being was to be defined, that was something that now and then I thought of dealing with, but it was something from which I always shrank and turned away. Finally, I tackled that subject in "That Thou Art Mindful of Him" (*Fantasy and Science Fiction,* May 1974) and full-circled myself back into the Frankenstein complex.

It may have been partly in expiation for this that I went on to write "The Bicentennial Man." There I considered not only what a human being might be but what a robot might be, too, and ended by showing, in a way, their coalescence.

12. In "That Thou Art Mindful of Him," then, you forecast the replacement of human beings by robots, while in "The Bicentennial Man" you forecast the fusion of human beings and robots. Which do you think is the more likely of the two?

Perhaps neither.

I feel that not all intelligence need be equivalent. Suppose dolphins have intelligence that is comparable to ours, as some people think. Its evolution and its way of life are nevertheless so different from ours that we seem to be able to meet on no common ground. Our respective intelligences seem to be so different in quality that there is no way of judging whether the dolphin is less advanced than we are or, possibly, more advanced, because there is no way of comparing them quantitatively.

If that is true in comparing the human being with the dolphin, how much more so might it not be in comparing the human being with the robot.

The human intelligence is the result of over three billion years of biological evolution, working through the processes of random mutation and of natural selection acting on systems of nucleic acids and proteins, with its driving criterion of success that of survival to the stage of reproduction.

The robot intelligence is the result, so far, of thirty years of technological evolution, working through directed human design and experiment acting on systems of metal and electricity, with its driving criterion of success that of usefulness for human purposes.

It would be very odd indeed, with every point so different in the two varieties of intelligence, if they did not end up very different—so different that no direct comparison is possible.

Robot intelligence seems to specialize in the scrutiny of tiny parts subjected to definite and repeated arithmetical operations with faultless accuracy and incredible speed. In that respect it far outmatches us already and may forever do so.

Human intelligence seems to specialize in an intuitive understanding of the whole and advances by the conjectural leap. In this respect we far outmatch the robot and may forever do so. After all, how can we program a robot to be intuitive if we do not know what it is that makes *us* intuitive?

Even if we could make the robot more like a human being, or vice versa, why should we want to? Why not take advantage of each area of specialization and make the robot ever better in its weighing of parts and the human being (through genetic engineering, eventually) ever better in its weighing of the whole?

We could then have a symbiotic arrangement, one in which the robot and the human being would be far greater together, than either could possibly be separately.

It was this which I was aiming at in my Lije Baley novels. *The Caves of Steel* (Doubleday, 1953) pictured a society in which human beings overbalanced the robots; *The Naked Sun,* one in which the robots overbalanced the human beings. The projected third novel of the series was to show the symbiotic balance—but though I tried I lacked the ability to picture what was dimly in my mind.

I failed when I first tried in 1958 and I never quite felt I was up to it since. What a pity I didn't get to it while I was still in my twenties and had not yet grown wise enough to know there were things I lacked the ability to do.

5* YES! WITH A BANG!

As a scientist, I like to think I can tell a suggestion that is scientifically untenable from one that is worth consideration, and I have no hesitation in dismissing the sort of silly stories we hear about flying saucers and Bermuda Triangles and pyramid power. There's not even any excitement or interest in that.

When there are suggestions that are scientifically tenable, however, but very dramatic and out-of-the-way, then all the science fiction writer in me comes to the fore. My eyes glitter and my breath quickens. And when there are two or more competing suggestions, all of which are dramatic, then I have no hesitation usually in picking the one I prefer—which is generally the one I consider most dramatic.

A little over a year ago, for instance, I wrote about the case of the mysterious layer of iridium-rich sediment in Italy that, on being tested for age, turned out to have been laid down just at the end of the Cretaceous, some 65 million years ago, when the dinosaurs died off (see "The Noblest Metal of Them All," in my collection *The Sun Shines Bright,* Doubleday, 1982).

That seemed like too much of a coincidence; surely there had to be a connection. Extraterrestrial matter, generally, is richer in iridium than the earth's crust is because on earth, most of the iridium collected in the core at earth's center. Might there have been some sort of splatter of extraterrestrial matter over the earth and might that have killed the dinosaurs?

Supernova? Meteorite? Solar explosion?

The supernova notion seems unlikely. It would have had to be pretty close to have had so drastic an effect and there isn't really any observation in the heavens that is consistent with a huge supernova only 65 million years ago. Besides, if the iridium had been of supernovan origin it would have had an abnormal ratio of isotopes, and that abnormal ratio was not found. In addition, there would have been plutonium-244, which would have been formed in a supernova explosion and which has a half-life long enough to have been still present in easily detectable quantities in that layer.

The possibility of a meteorite seemed to be thrown out by the fact that there was no sign of any collision at the site at which the iridium was located. Besides (I thought) even if a meteorite hit what is now Italy, how could it devastate the earth and kill dinosaurs thousands of kilometers away?

So I chose the solar-explosion hypothesis enthusiastically. After all, what with Maunder minima and neutrino deficiencies the sun has begun to seem frighteningly unstable in mysterious ways. Even a small blow-off, one that would scarcely affect the sun, would be enough to bathe us with solar material and produce a large enough wash of heat to wreak havoc with its life forms.

What killed off the dinosaurs then (I decided) was not the bang of a giant star exploding, or even the smaller bang of a meteorite smashing into the earth; what killed them was the whimper of a solar hiccup.

I chose the solar explanation, furthermore, because in a way it was the most frightening. While it was perfectly possible that a supernova might have exploded near us in the far past or a large meteorite might have hit us then, there is every reason to suppose that no star in our neighborhood can possibly explode during the next few million years and little reason to expect another large meteoric impact in the near future.

On the other hand, with our present uncertainty concerning the nature of the solar interior, how are we to know that there won't be another solar hiccup tomorrow?

Well, I seem to have been wrong, I am glad to say. The possibility of a near-sterilization of the earth tomorrow may appeal to my dramatic instincts as a fiction writer, but I don't really want it to happen in actuality.

As it happens, the evidence in the last year or so has rapidly mounted in favor of the meteoric hypothesis, which I had considered the least likely of the three.

I said, in my article last year, "I myself would like to see a thoroughgoing analysis of 65-million-year-old rocks in many places on earth, for a solar explosion would have affected the entire surface, it seems to me."

Well, that's been done. There have been analyses done in Denmark and in other areas of Europe; also in the north Pacific and in New Zealand, and the anomalous rise in iridium occurs everywhere and always in that same layer, the one that was laid down at the end of the Cretaceous.

I had pointed out that this would be in favor of the solar hypothesis, but, of course, it would be in favor of the supernova hypothesis as well, since that, too, would have affected the entire surface. What I didn't realize was that it would also support the meteoric hypothesis under certain circumstances.

I pointed out in the earlier article that whatever the cause, "it should also have resulted in raised values for some elements other than iridium."

That, too, has been tested, and it is found that there are raised values for such metals as osmium, palladium, nickel, and gold, for instance, in addition to iridium. As it happens, the relative concentration of these metals is just about that which is found in typical meteorites.

Opinion therefore began to swing markedly in favor of a meteorite impact. Since the effect seems to have been worldwide, there seems to have been an impact of a meteorite so large that it could blow itself into fine dust with such force that it would layer the whole earth with itself and not just the region surrounding the strike. That was why there were no signs of any impact in Italy—because that is not where it happened.

And that is where I fell short. In thinking about the meteorite I had utterly failed to consider the possibility of something sufficiently large scale. And I a science fiction writer!

What's more, I really mean large scale. In order to produce the effects it did, the meteorite would have had to be 10 kilometers (6 miles) across. It would not be just a meteorite. It would be an asteroid.

But where would so large an object come from?

From the space around us, of course. Such objects do exist there. So-called earth-grazers can approach disturbingly close to earth's orbit. Most of these earth-grazers are only a kilometer or so across, and though that would be enough to wreak havoc on the human scale, it would not be enough to bring about a near-sterilization of the planet.

There are occasional larger ones, too, however. Eros, the largest of the earth-grazers and the first to have been discovered (in 1898), is 24 kilometers (15 miles) across at its longest diameter. We seem to be safe from it, though, since its orbit is fully 22.5 million kilometers (14 million miles) from earth's orbit at the point of closest approach, but there may have been an Eros-like body that could come closer than that and that we're not aware of only because it no longer exists—having destroyed itself against the earth 65 million years ago.

Yes, but any earth-grazer with an orbit that could send it crashing into earth would surely have done so long ago, much longer than 65 million years ago. Once the dangerous ones had been swept up—billions of years ago—wouldn't space be clear? Isn't that an argument against the meteorite hypothesis so short a time ago?

That would be true if earth-grazers had orbits that never altered. That, however, is not so and can't be so. An earth-grazer inhabits the inner solar system, and every once in a while it passes relatively close by one of the large bodies of the inner solar system: Mars, earth, moon, Venus, or Mercury. Each time it does so its motion must be affected by the large body's gravitational influence. The earth-grazer is then perturbed into a new orbit, the change being a very slight one, if the distance between itself and the perturbing body is great; pronounced, if the distance is small.

(To be sure, the earth-grazers also produce perturbations in the large bodies, but these perturbations are in inverse proportion to the relative masses and since the large bodies are billions of times as massive as the earth-grazers, the perturbations of those large bodies are insignificant.)

The result of the perturbations is that over comparatively long periods of time, every earth-grazer wanders rather widely over the inner solar system and, sooner or later, is sure to take up an orbit that intersects that of earth.

Once that happens, a collision is bound to take place before very long, astronomically speaking, unless another perturbation moves the orbit in such a way as to make that collision impossible.

In the long run, then, not only will there be collisions, but the incidence of such collisions will not decrease markedly with time.

To be sure, every earth-grazer that hits the earth (or one of the other large bodies in the inner solar system) is one earth-grazer less. In addition, every once in a while an earth-grazer is perturbed in such a way that it will adopt an orbit that will carry it out of the solar system altogether. Balancing that, however, is the fact that an asteroid that is not an earth-grazer is sometimes perturbed into becoming one, so that new dangers periodically arise.

Actually, earth suffers an endless series of collisions and what saves it is the fact that virtually all the collisions are with tiny bodies. This is not because the earth has a special affinity for tiny bodies, but only because there are more smaller bodies than larger bodies in any class of astronomical objects, and that includes earth-grazers.

Thus, the number of dust-sized particles that hit the earth—or at least enter its atmosphere—and float slowly downward as meteoric dust is in the trillions per day. Those particles, with sizes up to pinheads, that are large enough to be heated to a white-hot flash but are not large enough for any part to survive the atmospheric passage and hit the ground as anything more than dust are fewer but still in the millions.

The number of objects the size of pebbles and rocks that are large enough to survive the flight through the atmosphere and reach the ground as meteorites is smaller still; perhaps one a year the world over. And of these, the larger the meteorite we're considering, the longer the interval between strikes.

An earth-grazer, 10 kilometers (6 miles) across, might be expected to hit the earth every 100 million years on the average. (Averages are tricky things, of course, and don't represent a strict schedule. There's a tiny chance that two might hit in successive years, and a tiny chance that earth might not have been hit at all in the course of its lifetime. The chances are, however, that once the earth settled down, 4 billion years ago, after having swept up most of the loose matter in its orbit, it was struck some forty times with good-sized earth-grazers.)

I suppose astronomers had reason for thinking in this way ever since Eros was first discovered, and better reason for doing so with each passing decade as more and more earth-grazers were discovered, and as the orbits of those that were detected were found, in some cases, to make disturbingly close approaches to earth's orbit.

Then, too, the sudden ending of the dinosaurs was an event sufficiently dramatic to lure some scientists into suspecting a catastrophe. There have been other episodes of "Great Dyings," but the one that took place 65 million years ago was not only the most recent, and therefore the best documented in the fossil record, but the one that involved the most spectacular creatures—the largest and most magnificent land animals ever to dominate the earth.

At any rate, in 1973, without any real evidence, Harold C. Urey suggested that it was a cometary impact with the earth that had ended the dinosaurs.

Most scientists, however tempted, did not wish to choose a catastrophic solution. For one thing they suspected that the "sudden" ending of the dinosaurs might not have been as sudden as all that. They might have died off over the space of a few hundred thousand years, and to the fossil record that might have seemed a sharp ending. In that case, it would be more fruitful, perhaps, to look for some slow change in earth's environment which at some point set the dominoes in motion. A lowering in temperature, or a raising of ocean salinity, a draining of shallow seas, a mountain upheaval . . .

It was not until the coming of the iridium anomaly that the catastrophic solution began to look too good to dismiss.

But why should an earth-grazer impact kill off the dinosaurs?

It was worse than that, in fact, for it was not only the dinosaurs that died off. Other spectacular reptiles—the plesiosaurs, the ichthyosaurs, the pterosaurs—died off simultaneously. So did the invertebrate ammonites. So did a wide variety of microscopic creatures.

How could all that have happened?

Suppose we imagine an earth-grazer, 10 kilometers (6 miles) across, swooping down toward earth. It must have made a huge flash and the great-grandfather of all thunderclaps as it struck. The noise was probably heard all over the world, and if that was so then the dinosaurs went out, yes, with a bang!

It would mean that about 1.5 trillion tons of matter was striking the earth at a speed of about 25 kilometers per second. The enormous kinetic energy of that strike would have reduced the earth-grazer and the surrounding regions of earth's crust to dust and vapor and would have thrown into the stratosphere a quantity of dust equal to ten or twenty thousand times the mass of the earth-grazer itself.

All of it would eventually have settled back to earth, and that portion of the dust that was the earth-grazer would have been responsible for the thin layer of greater-than-earthly-normal concentrations of iridium, osmium, palladium, and so on.

The dust, however, would not settle back immediately. It takes time. The volcanic explosions of Krakatoa in 1883 and of Tambora in 1815 spewed dust into the stratosphere that remained there for a couple of years in sufficient quantity to produce noticeable effects. Tambora, which produced the greater supply of dust, reflected enough sunlight away from earth by means of that supply to produce the "year without a summer" in 1816.

Tambora, however, delivered half a trillion tons of dust into the atmosphere at most, while the earth-grazer strike at the end of the Cretaceous delivered at least forty thousand times that quantity. If Tambora could produce a year without a summer, what would the earth-grazer strike do?

The estimates are that the dust produced by that enormous strike would fill the stratosphere to such an extent that sunlight would simply not get through.

After the strike had taken place, and the atmosphere had stopped shaking with the sound, and the earth itself had stopped ringing with the blow, and the volcanoes, earthquakes, tsunamis, had all done their worst, there came something far more deadly still—a creeping darkness.

I don't know how long it would take the dust to spread out over the world, but suppose we consider some spot of land thousands of kilometers from the strike, and imagine some intelligent observer on the spot.

The observer would have heard the distant sound and might have been removed from any immediate consequences, but he would notice that the sun was beginning to be ruddy as it rose and was staying ruddy as it climbed higher in the sky. Each day it would rise redder and dimmer, and each day would be rather colder than the one before. And finally, one morning the sun would not truly rise at all. There would be a dim lightening of a black sky and that would be all, day after day after day.

It is estimated that at the peak of the dust cover, perhaps no more than one 5-millionth of the sun's light penetrated the dust layer so that the earth was bathed in a diffuse light only one-tenth as bright as that of the full moon. Everything else in the sky, the moon, the planets, and the stars, was totally gone. And this may have continued, with slowly decreasing intensity, for up to three years!

In that long winter, that long dark winter, earth's plant world died, and because the plants died, the animal herbivores died, and because the herbivores died, the animal carnivores died. That included the dinosaurs, of course, whose end was sudden indeed, for all the flourishing lot of them must have died in the course of three years.

Of course, this new view of an earth-grazer strike brings problems. It easily explains the extinction of three-quarters of the species that existed on earth at the end of the Cretaceous. The problem is to explain how the other quarter managed to survive, even if only in greatly reduced numbers. Why wasn't the earth utterly sterilized?

We might reason that spores, seeds, and root systems lay dormant through the long dark, and then as the dust cover began to thin and the sun's disk began to brighten and the warmth began to steal back to earth's surface, they quickened. Plankton began to reappear in the ocean while bits of green began to touch the desolate land once more.

Little by little, the hold of plant life on the planet strengthened until the world was smiling and warm and green again—but was populated only by survivors.

There were some animals that had managed to eke out a spare living on the remnants of dead life, on seeds, on frozen carcasses, and they revived, too, until the earth was once again overrun with animal life—again survivors. The huge clumping tread of the grand dinosaurs was gone, to be replaced by the scurrying patter of small mammals and the whirring flight of small birds.

It sounds good, but paleontologists are going to have to explain exactly why certain species died and other species survived, and whatever scenario they devise, it is sure that someone will say: "Then why did species A survive when it couldn't possibly have taken advantage of those methods of survival?" or "Why did species B not survive in that case?"

Some paleontologists are so daunted by the difficulties of accounting for survival and nonsurvival that they don't think it can be done by anything as heavy-handed as an earth-grazer strike and a massive dust layer in the stratosphere. They want to make use of changes that are slower, more selective, and less dramatic. In that case, though, they're going to have to explain not only the pattern of extinctions, but also that worldwide layer of iridium-high sediment.

That won't be easy, either.

But if an earth-grazer struck the earth 65 million years ago with enough force to send all that dust into the stratosphere it must have left a memento behind in the form of a crater gouged out of the earth's crust. That crater would have to be 175 kilometers (110 miles) across and have an area equal to that of the state of New Hampshire. It should be noticeable—so where is it?

To be sure, the 65 million years that have passed since the crater formed gives ample time for the action of wind, water, and life to erode it pretty much to nothingness so that it would vanish from ordinary observation. But not entirely.

There would be left circular structures still visible from the air—a circular disruption of the rock formations, possibly a circular lake.

There are such formations here and there on earth. The Great Meteor Crater in Arizona is the most obvious example—but it is quite small and was formed only some tens of thousands of years ago. In eastern Quebec, however, there is a circular lake-filled structure some 70 kilometers (43 miles) wide and perhaps 210 million years old.

If that Quebec formation can still be seen, then the strike that ended the Cretaceous, producing a circular area more than six times as great and only a third as old, should certainly be seen. Where is it?

No problem. The chances are seven to three that the earth-grazer struck the ocean, and the ocean it must have been. It went screaming through the water in a matter of seconds, with water hissing and boiling about it, and gouged out the sea bottom, sending up a cloud of water vapor along with the dust and debris.

The crater, then, would be somewhere on the sea bottom if the shifting tectonic plates of the earth's crust have not obliterated it, and we may yet discover traces of it as we explore the sea bottom in detail. If we do find it, that would be extraordinarily good evidence in favor of the strike and also the best excuse yet to send a bathyscaphe down for a detailed study of a specific underwater formation.

Of course, a sea strike would mean a tsunami—a huge splash of water. All the islands and continents of the world would receive that incredible wash of ocean, some more than others, of course. That, too, would have contributed to the devastation of land life and should be taken into account in reckoning extinctions.

But here I must include a speculation by my good friend the astronomer Fred Whipple. He sent me an advance copy of a paper entitled "Where Did the Cretaceous/Tertiary Asteroid Fall?", the contents of which I have his permission to mention.

He suggested that the earth-grazer might have happened to strike near the junction of two of the plates that make up the crust of the earth. He estimates that, as a result of a random strike, there is one chance in twenty-five that it would land within 200 kilometers (125 miles) of such a plate junction. This is not a good chance, but not a very farfetched one either.

If a strike had been made at or near a point of junction, it would have produced a much more effective puncture of the crust, a much more prolonged period of volcanic activity, and a much greater effusion of new land surface. The new land would have wiped out any crater, but the new land itself should represent a recognizable phenomenon.

Whipple asks if there is any sizable hunk of relatively recently formed land area sitting astride a plate junction, and he points out that there is one and only one land area that seems to fit all the requirements, and that is the island of Iceland, which lies astride the junction of the Eurasian and North American plates.

Whipple further points out that whereas the Italian sediments, where the iridium anomaly was first detected, showed thirty times the normal values, similar analyses in Denmark showed 160 times the normal values. But then, Denmark is closer to Iceland than Italy is (even allowing for plate movement over the last 65 million years) and we can suppose that some portion of the earth-grazer material thrown up was in pieces large enough to settle out comparatively rapidly so that Denmark received more fallout than Italy did. New Zealand, on the other hand, which was much farther away than Italy, has an iridium content only twenty times greater than normal.

Iceland's mere existence is an interesting point in favor of the earth-grazer strike, and if the amount of iridium continues to decrease with distance from Iceland as more and more areas are studied, then I think it will be impossible to refute the earth-grazer explanation of the end of the Cretaceous, along with Fred Whipple's refinement thereof.

One last point of my own. The readers of this essay series must know I am a great collector of historical coincidences. I find it interesting, therefore, that there is mention in world literature of severe winters over a period of years that led to the end of the world as it was then known and the reconstitution of a new world.

That was the Fimbul winter ("terrible winter") spoken of in the Norse myths. It lasted for years and was the prelude to the final battle of Ragnarok between the gods and the giants in which the world as it existed was destroyed.

And what is the source of our knowledge of Fimbul winter? Why, the writings of Snorri Sturluson (1178–1241) of Iceland. It follows that the tale of the mythic Fimbul winter comes from the very place where the strike took place that started the real Fimbul winter.

Any connection? A racial memory? Some mystic lingering?

Of course not. Not after 65 million years. There's a much more natural explanation.

Considering the climate of Scandinavia, it would be logical to suppose that the end of the world would be heralded by severe winters. In warmer climates, severe droughts would have been called on for the purpose.

And, as for Iceland, that happened to be, through the accident of geography, the last area of Scandinavian culture to fall under the influence of Christianity, so that the pagan legends survived best there.

But it's a nice coincidence.

6★ LET ME COUNT THE DAYS

I have had published a revision of *Asimov's Biographical Encyclopedia of Science and Technology*. When I first published the book in 1964, it contained the short biographies of about a thousand scientists (including a few inventors and explorers) arranged in chronological order of birth. In the second edition, published in 1972, I revised and lengthened many of the biographies and increased the number to 1,195. In the third edition, published in 1982, there was again a revision and lengthening of many of the biographies and a further extension to the number of 1,510.

For each biography I begin by giving the place and the date of birth and death (to the exact town and day, when I can find it) and in doing so have come across a peculiar fact. During my work on the first edition, I couldn't help but notice that a large majority of the famous scientists of all periods with whom I dealt had lived long lives and had died at ages well beyond my own. Somehow that is no longer the case. I can't explain it, but in the third edition, a surprisingly large fraction of the scientists I deal with have died prematurely at ages *less* than my own.

This got me to thinking about the age at death of my scientists, and for a while I toyed with the idea of adding that age to the initial statistics.

I decided against it for it seemed to me that it might be a touch ghoulish to do so. Besides, given the impossible calendrical system that humanity has worked out, it would take me a period of time to calculate the exact age. Not much time for any one entry, but multiply that by at least a thousand where the exact birth date and death date are known and it would come to a substantial investment of time and one that I was reluctant to make.

For instance, the longest-lived scientist in the book is Michel Eugène Chevreul, a French chemist who was born in Angers on August 31, 1786, and who died in Paris on April 9, 1889, so that when he was a child he could watch aristocrats on the way to the guillotine during the Reign of Terror and when he was an old, old man he could watch the Eiffel Tower under construction.

How old was he when he died? Well, if you simply subtract the year of birth from the year of death, you get 1889 — 1786 = 103. However, he was born in August and died in April, so he was a bit short of his 103rd birthday when he passed away. To be exact, he was 102 years, seven months, and ten days old when his life was done. To be still more exact, and allowing for the varying lengths of months and the precise intervention of leap years, Chevreul died on the 37,476th day of his life. (We can ignore the matter of hours, minutes, and seconds, since my statistical information is never that precise.)

Here's the way that is calculated. From August 31, 1786, to August 31, 1888, we have 102 years, which, allowing 365 days per year, comes to 37,230 days. There is a leap day every four years; the first in Chevreul's life coming on February 29, 1788, and the last on February 29, 1888. By the Gregorian calendar, however, there was one interruption in the every-fourth-year cycle, since 1800 was *not* a leap year. Omitting 1800, that means there were 25 leap days in the course of Chevreul's life, and these must be added to the number of his days, and that brings us to 37,255. The period from August 31, 1888, to April 9, 1889, the actual day of Chevreul's death (allowing 30 days for September and November, 31 days for October, December, January and March, 28 days for February, and 9 days for the fragment of April) gives us an additional 221 days which brings us, as aforesaid, to a total of 37,476 days.

As you see, this sort of thing would grow tedious indeed if I tried to work it out for every person in my book for whom I had the exact birth date and death date.

If, instead of counting days, months, and years as three separate but interconnected systems, people had merely counted the days, then the task would have been much simpler. A single subtraction per person would at once give the exact age at death. Still ghoulish, you understand, but it would then have been an irresistible task for anyone with my own overwhelming compulsion to count things.

As it happens, though, astronomers *do* count the days. They do so according to a system invented in 1583 by a man named Joseph Justus Scaliger.

Scaliger was a French scholar who was born in a small French town named Agen on August 5, 1540, and died in Leiden, the Netherlands, on January 21, 1609, on the 25,005th day of his life, so that he had a lifetime almost exactly two-thirds as long as that of Chevreul.

Scaliger was inhumanly driven by a scholarly father into an encyclopedic knowledge of the classical authors. The treatment was not as cruel as it might have been, for the young Scaliger took to his tasks with extreme avidity and apparently loved to pile up knowledge of all kinds.

He was converted to Protestantism in 1562 when France was badly divided between the Catholics and the Protestant Huguenots. The two parties were engaged in a long-drawn-out series of civil wars in which the Huguenots, rather surprisingly, held their own despite the fact that they made up only 10 percent of the population.

The low point for the Huguenots came on August 23, 1572 (the day of the year which was dedicated to St. Bartholomew in the Church calendar). The Catholic party took advantage of a truce to attack the unprepared Huguenots and to kill some tens of thousands of them in cold blood. Scaliger, either by a fortunate chance or by keen foresight, left Paris just before this "St. Bartholomew's Day Massacre" and went to the Protestant stronghold of Geneva, Switzerland, so he survived. In 1593, he took a professorial position with the University of Leiden (again a Protestant center) and remained there till his death.

He died in humiliation, however. Scaliger's father had always maintained that he was connected by birth to the old ruling family of the city of Verona in Italy and young Scaliger believed that implicitly. He proudly spread the news every chance he had and in all controversies he used his own noble blood as a bludgeon against his contemptible lowborn opponents.

Toward the end of his life, however, evidence surfaced which made it perfectly clear that the elder Scaliger had lied. Scaliger's enemies jeered and Scaliger wilted and died.

Scaliger was particularly interested in chronology. He considered not only Greek and Roman history, but studied every scrap of record he could find of the various barbarian nations. He considered not only the biblical record of the Jews but everything that existed concerning the histories of Egypt, Assyria, Chaldea, Persia, and so on.

Naturally, every land had its own system of keeping track of time and Scaliger managed to trace out no fewer than fifty different calendars. He did his level best to match them, one against the other, so that he could convert the time when an event took place according to one calendar into what the time would have been by any other of the calendars he studied and, particularly, what it would have been in the calendar that was being used in western Europe at the time.

In this way, he could produce the outline of a world history in which all events everywhere could be put into one well-defined order. One would know just what was happening in Greece when something else was happening in Persia; or what was happening in Carthage or in Gaul when something was happening in Rome and so on.

In point of fact, Scaliger is considered the founder of modern chronological studies.

Scaliger saw that it was a wearisome task to deal with calendars in general, even the best of them, because all were beset with the notions of days and weeks and months and seasons and years in an unending mishmash. The only way to cut completely through all calendrical confusion was simply to count the days. The day was common to all calendars and it was unthinkable that there could be a culture anywhere that would not consider the day a natural unit of time. The overwhelming fact of sunrise and sunset could not possibly be ignored (except in the polar regions, of course, but no calendrical system reached Scaliger from the thin leaven of Eskimos who inhabited the shores of the Arctic Ocean).

Scaliger therefore set about numbering the days according to what he called the *Julian period*. It is usually thought that he was in this way honoring his slave-driving, birth-bragging father, Julius Caesar

Scaliger. It may also be, however, that he named it for the Julian year, which was fundamental to his calculations, and which was named for another Julius Caesar—*the* Julius Caesar, in fact.

It may seem that it is a simple task to number the days. You just go 1, 2, 3, 4, and so on as long as the days hold out. You will never run out of numbers.

The question is, though, which is day 1?

You might let *today,* the day on which you think of doing the numbers, be day 1. Then tomorrow is day 2, and so on.

Perhaps it would be more impressive, however, to think up a particular date in the past, some date on which something of great significance to you or to your culture had taken place, and let that be day 1. You might pick your birth date as day 1, or you might (if you are an American) decide to make July 4, 1776, the birth date of the nation, day 1.

Scaliger might have chosen his own birth date, or, perhaps October 31, 1517. That was the day on which Martin Luther nailed his list of ninety-five theses on which he was ready to dispute to the door of the Wittenberg church, thus beginning the Protestant movement. Or he might have picked the birth date of Jesus as day 1.

Scaliger, however, was a chronologist and he wanted a day 1 that had calendrical significance. For instance, it had to be on some January 1, so that day 1 would begin a year. But which year?

Well, there were two kinds of years that were used in classical times. There was the solar year based on the movements of the sun in the sky, and that was 365 days long. It was invented by the Egyptians and a slightly improved version of it was introduced into Rome by Julius Caesar in 45 B.C. It was called the Julian year in his honor.

In the Julian year, the year is taken to be 365.25 days long, so that four years are exactly 1,461 days long. In order to make each year contain a whole number of days, you arrange to have three years of 365 days each and then a fourth year of 366 days; this pattern repeating itself with metronomelike regularity.

The advantage of the solar year is that it keeps time with the seasons so that planting time, harvest time, wet and dry, winter and summer, come at the same region of the calendar year after year.

There is also a lunar year based on the phases of the moon. This was invented by the Sumerians and was inherited by the later peoples of Babylonia. It spread to the Jews and the Greeks and it still forms the basis of the Jewish religious calendar. It is also used by the Christian world, even today, to calculate the day of Easter.

In the lunar calendar, there are twelve lunar months (the time from new moon to new moon) in the year. A lunar month is about 29.5 days long (29.53059 to be exact), so that lunar months can be made alternately 29 and 30 days long. However, twelve lunar months come to 354.367 days, which is about 11 days short of the length of the solar year. This means the lunar calendar will quickly get out of phase with the seasons unless that 11-day discrepancy is made up.

If you were to add the 11 days each year, you would throw the year out of synchronization with the phases of the moon, which would go against religious tradition. In those societies with a lunar calendar, people therefore waited till the shortage amounted to about a whole lunar month and then added one. This meant that every third year or so there would have to be a year with 13 lunar months.

Somewhere about 500 B.C. it struck Babylonian astronomers that a period of 19 solar years was just equal to a period of 235 lunar months (235.003, to be exact, but let's not quibble).

With 235 lunar months, one could have 12 lunar years of 12 months each and 7 lunar years of 13 months each. A cycle of 19 years was set up with the 13-month years spread more or less evenly among them—making up the 3rd, 6th, 8th, 11th, 14th, 17th, and 19th years, actually. At the end of the cycle the lunar calendar would be exactly even with the sun and a new cycle could begin.

In other words, if January 1 of a certain year is blessed with a new moon (so that it would be the start of the year in both the lunar calendar *and* solar calendar) so would the January 1 that was 19 years earlier, and 19 years earlier than that and so on indefinitely.

It seemed to Scaliger that day 1 ought to be on a January 1 with a new moon so it could start both years.

As it happened, January 1, 1577, fell on the new moon. If we count back 83 lunar cycles of 19 years each, we count back 83 × 19, or 1,577 years. This brings us back to the year 1577—1577, or 0. But there is no year 0. The year before A.D. 1 is 1 B.C. Therefore, it turns out that January 1, 1 B.C. was on the day of the new moon.

Scaliger may have had the impulse to make that day 1, for it was his belief that Jesus was born on December 25, 1 B.C., and it must have seemed significant that January 1 of the year of Jesus' birth began both a solar year and a lunar year. (Actually, Jesus couldn't have been born in 1 B.C. because, according to the circumstantial account in Matthew, Herod the Great was alive at the time of the birth of Jesus and Herod died in 4 B.C. Jesus must have been born at least one or two years earlier than that—but no one really knows.)

Scaliger resisted the impulse to make January 1, 1 B.C., into day 1, however. To have chosen any particular date that is embedded in history as day 1 would have created unnecessary difficulties.

For instance, by a convention made official only in the time of Charlemagne, a misplaced piety has counted the years from the birth of Jesus. This means that we have a double system of counting—forward from A.D. 1 and backward from 1 B.C. with no year 0 in between (because Europeans knew nothing about the zero symbol in Charlemagne's time). This introduces needless confusion and complication in chronological studies.

Indeed many unsophisticated people find the whole notion of "B.C." too complicated to consider and wipe it out. I suspect that millions in the United States have a vague feeling that history started less than two thousand years ago. The common saying "ever since the year 1" makes it seem as though time began with A.D. 1.

Before the introduction of this "Christian era," many people used a "mundane era" of one variety or another, by which the years were counted from the time of the biblical creation of the world. There was no general agreement, however, on exactly when the world was created since the Bible has no chronological system, from beginning to end, that anyone can make indisputable sense of.

Nevertheless, the date generally accepted by Protestant fundamentalists today is 4004 B.C.

If we were to consider January 1, 4004 B.C., as day 1, that would be rather convenient for it would eliminate any need for negatively numbered days. Though the earth existed for billions of years before 4004 B.C. (something no one dreamed of in Scaliger's time), written history did not. All the events of historical times for which one could reasonably try to find an exact date have taken place after 4004 B.C. and would therefore have fallen on a positively numbered day.

Scaliger, however, did *not* choose January 1, 4004 B.C., as his day 1 because, for one thing, that date was worked out by the Anglican bishop James Ussher about 1650, which was over forty years after Scaliger's death. Nevertheless, all the mundane eras in use placed the creation of the world sometime before 3500 B.C., and Scaliger was chronologer enough to know he needed a date at least that early for his day 1.

To find some significant early date, Scaliger considered the solar cycle. This has nothing to do with the sun as an astronomical body. It deals, instead, with the day of the week on which January 1 of a particular year falls. The arrangement is neatest when January 1 falls on a Sunday so that it begins the week as well as the year. In Latin, Sunday is *dies solis* and that is why it is referred to as the "solar cycle."

Since most years are 365 days long, they are 52 weeks and 1 day long. That extra day means that if January 1 is on a Sunday in a given year, it falls on a Monday the next year, on a Tuesday the one after and so on. If all years were 365 days long, January 1 would fall on a Sunday every 7 years.

In the Julian calendar, however, every fourth year is 366 days long, or 52 weeks and 2 days. If January 1 of a 366-day year falls on a Sunday, the January 1 of the next year is on a Tuesday. It "leaps over" Monday, so that a 366-day year is called *leap year*.

Suppose, then, we number the days of the week from Sunday to Saturday, 1 through 7, and imagine January 1 of a leap year falling on Sunday (1). The next year January 1 will fall on Tuesday (3), then the next year on Wednesday (4), then Thursday (5), then Friday (6). But this year is another leap year so that January 1 on the year after skips over Saturday (7) and is on Sunday again (1).

In fact, the progression goes like this if we place an asterisk to mark the leap years: 1*, 3, 4, 5, 6*, 1, 2, 3, 4*, 6, 7, 1, 2*, 4, 5, 6, 7*, 2, 3, 4, 5*, 7, 1, 2, 3*, 5, 6, 7, 1*.

If you count, you will see that 28 years after a January 1 of a leap year that falls on a Sunday, there will be another leap year in which January 1 will fall on a Sunday, and everything will then repeat over and over. (We don't really have to count. There are 7 days in a week and 4 years to a leap year, and 7 and 4 are mutually prime—there is no number other than 1 that will divide both 7 and 4 exactly—so that the length of the cycle is 7×4, or 28.)

As it happens, 1560 was a leap year (all years with numbers which are divisible by 4 are leap years in the Julian calendar—at least in the A.D. part of it). If we count backward in 28-year intervals, then 56 such solar cycles take us back to 9 B.C., the nearest such year to the birth of Jesus. (The year 9 B.C. is a leap year even though 9 is not evenly divisible by 4, because there is no year 0. In the B.C. portion of the calendar, leap years have numbers which, when divided by 4, leave a remainder of 1. I'm sorry about that but that's what comes of leaving out year 0.)

Had 1 B.C. (a leap year, by the way) had January 1 on a Sunday, so that that January 1 began not only the lunar year and the solar year but the week also, Scaliger might not have been able to resist calling it day 1. Fortunately, it fell on a Wednesday and Scaliger had to look further.

Are there years in which the solar cycle and the lunar cycle coincide and where January 1 begins the week, the lunar year and the solar year; in other words, in which January 1 of a leap year falls on a Sunday with a new moon in the sky?

Certainly! Since the solar cycle is 28 years long and the lunar cycle is 19 years long and 28 and 19 are mutually prime, the combined cycle is 28 × 19, or 532 years long. In other words, if January 1 of a certain leap year falls on a Sunday with a new moon, then 532 years earlier there was another such January 1 and so on. This 532-year cycle was first pointed out by Victorinus of Aquitaine about A.D. 465.

The year 1140 was a leap year in which January 1 fell on a Sunday with a new moon, so this was also true for January 1, 608, for January 1, 76, for January 1, 457 B.C., and so on.

One of those dates, if one counted back long enough, would be suitable for day 1, but which one? Scaliger didn't want to choose arbitrarily. He needed additional significance. He therefore took into account the period of the indiction.

The indiction is the year in which, by Roman law, a census was taken of property and individuals in order to set up some system of taxation. The Emperor Diocletian, about A.D. 300, decreed that the indiction should take place every 15 years, and the custom survived the fall of the empire and right into Scaliger's time.

The period of the indiction had no astronomical significance at all; it was entirely man-made; but Scaliger may have picked it because its length, 15 years, introduced a number that was mutually prime with the 532 years of the combined solar-and-lunar cycle. The factors of 15 are 3 and 5, and those of 532 are 2, 2, 7, and 19, so that no particular factor appears in both. That means that the combined solar-lunar-indiction cycle, or "Julian cycle," is 532 × 15, or 7,980 years long.

Scaliger counted far back to find a leap year that was also an indiction year and in which January 1 fell on a Sunday with a new moon. That year turned out to be 4713 B.C.

That year was long, long before there was any indiction, or any Roman Empire, or any Julian calendar—but that didn't matter. Scaliger had what he wanted and he allowed January 1, 4713 B.C., to be his day 1. This manner of counting gives each day thereafter its own "Julian day."

Scaliger published his investigation of chronology, including his introduction of the system of Julian days, in 1583, and, as luck would have it, the very basis on which he founded the system, the Julian Year, came to the beginning of its end the year before.

In 1582, Pope Gregory XIII formally recognized that the year was not 365.25 days long but 365.2422 days long and decreed that ten days, which had been wrongly accumulated over a period of over a thousand years through the use of a slightly overlong year, be dropped. As a result the day after October 4, 1582, was October 15 and not October 5.

To keep further accumulation from taking place, three leap years were to be dropped every four centuries. In other words, instead of having every single century year (1600, 1700, 1800, etc.) a leap year, only those century years divisible by 400 would be leap years. That is, 1600 and 2000 would be leap years, but 1700, 1800, and 1900 would not be.

Protestants did not at once recognize this new "Gregorian calendar" since it was promulgated by a pope, but little by little, they fell into line. Nowadays, the Gregorian calendar is global. All nations use it for international dealings, if nothing else.

Thus, the Julian cycle of 7,980 years is 2,914,695 days long so that Julian day 2,914,696 would fall on January 1, 3268, and perhaps start a new count. By the Gregorian calendar, however, Julian day 2,914,696 will fall on January 23, 3268, and the neat balance will be upset.

That doesn't matter, however. Whatever the course of thought that led Scaliger to choose January 1, 4713 B.C., it was an adequate choice and once chosen all the scaffolding of this period or that cycle can be torn away and discarded. It is only necessary to keep counting the days from January 1, 4713 B.C., without any regard to calendar rules and calendar reforms.

What's more, the Julian day is in actual use; it is not just the peculiar quirk of a nutty chronologer. Astronomers use it routinely and find it a great convenience. They start the Julian day at noon (to leave the night—when observations are made—unbroken by a day change) and assign each day its number.

Thus, Halley's comet reached its perihelion on a certain Julian day in 1835 and on another in 1910, and by merely subtracting the former from the latter, it can be seen that the period of Halley's comet was 27,183 days. Other cyclical manifestations, such as the periods of variable stars, can also be dealt with simply and profitably.

Converting a particular calendar day into a Julian day, or vice versa, using either a Julian calendar or a Gregorian calendar (or, in principle, any calendar with fixed and rational rules), can be carried through according to a straightforward system of calculations which are, to be sure, very tedious.

These days, however, it is only necessary to program a computer and it will grind out Julian days for any calendar day you wish in a twinkling.

As it happens, at noon on January 1, 1982, Julian day 2,444,971 started. Knowing that (if you have nothing better to do) you can work out the Julian day of your birth date and of the day on which you read this and at once know how many days you have lived. I'd do it for myself, but I don't think I want to know just how many days I've lived at this particular moment.

7★ COUNTING THE EONS

In the Book Review section of (as I write) today's Sunday New York *Times,* there is an article entitled "What the Chinese Are Reading," written by Lloyd Haft of the University of Leiden in the Netherlands.

I began to read the article with one question in my mind, which was (as you can all guess) "Are they reading me?"

My hopes rose when, at the end of the first page, it was stated that at a certain book fair in China, there was, on the blackboard, a chalked introduction to a particular branch of literature. It was entitled "What is 'Science Fiction'?" and the description was not bad at all. It was, however, in connection with a display of Jules Verne's *From the Earth to the Moon* and, as I read on, I found that modern science fiction was not mentioned, unless you want to count Saul Bellow's *Mr. Sammler's Planet.*

I kept on, rather dispirited. Toward the end of the article, mention was made of a new Chinese magazine, *Dushu* (*Reading*), which dealt with non-Chinese books. Then came a magic sentence: "Among the books recently reviewed were *Letters of D. H. Lawrence,* Isaac Asimov's *In Memory Yet Green* and Joseph Heller's *Good as Gold.*"

I was delighted. The Chinese had indeed heard of me.

And yet *In Memory Yet Green* is not one of my works of fiction, or even one of my ordinary works of nonfiction. It is the first volume of my autobiography, one in which I was incautiously frank. I couldn't help but realize that now everyone knew how old I was—even the Chinese. The undeniable fact is that I was born on January 2, 1920.

In fact, Doubleday made that uncomfortably evident, for when they published the book in 1979, they placed under the title the printed notice *The Autobiography of Isaac Asimov, 1920–1954,* and surrounded it with a black border. This must have led many people to say, "Poor man! Died at thirty-four! Who's been writing all those books under his name in the last quarter-century, I wonder?"

The second volume of my autobiography came out in 1980. It is entitled *In Joy Still Felt* and has the subtitle *The Autobiography of Isaac Asimov, 1954–1978*. Now people will say, "Poor man! Died at twenty-four! How did he manage to write all those books in his short lifetime?"

Well, I didn't die in 1978 either, you bunch of wise-guy kids, but if I've got to brood about having entered early middle age, I can make up for it by taking up the matter of objects even older than myself.

The earth, for instance.

How old is the earth? Prior to the eighteenth century, in our western tradition, people relied on the Bible, and from what data it gave us, most calculations seemed to make it six or seven thousand years old. The most familiar of these calculations is one made by Archbishop James Ussher of the Anglican Church, who, about 1650, calculated, on the basis of his biblical studies, that the creation of the earth took place at 8 P.M. on October 22, of 4004 B.C. (I don't think he specified Greenwich mean time. Perhaps he assumed the earth to be flat.)

The first person in our western tradition to attempt to probe back beyond this biblical limit was the French naturalist Georges Louis Leclerc de Buffon (1707–88), who, in 1745, dared to suggest that the earth was created not by the word of God but by the collision of a massive body (which he called a comet) with the sun. He guessed that this had happened 75,000 years before, and that life had come into being perhaps 40,000 years before. It was a very daring suggestion for its time and it got him into some trouble with the theologians. Fortunately for himself, Buffon was not one of your hard-line controversialists, but knew how to retreat gracefully in the face of Ignorance waving the Bible.*

Next came the Scottish geologist James Hutton, who, in 1795, published a book called *Theory of the Earth.* In it, he carefully described the slow changes taking place in the earth's crust today. It seemed clear that some rocks were laid down as sediment and then compressed into hardness; other rocks made their appearance as molten lava from earth's depths and then cooled into their solid shape; exposed rocks were worn down by wind and water. It all happened with excessive slowness.

Hutton's great intuitive addition to all this was the suggestion that the forces now slowly operating to change the earth's surface had been operating in the same way and at the same rate through all earth's past. This is the *uniformitarian principle.*

In the following decades, geologists began to try to use the uniformitarian principle to calculate the age of the earth, or at least of some geological phenomena. They made rough calculations as to how long it took to lay down an observed thickness of sedimentary rock; how long to form a particular river delta out of the mud carried downstream by the river. While the calculations were approximate at best and required a bit of guesswork here and there, it seemed quite obvious that to account for what existed on earth, its age would have to be not in the tens of thousands of years as Buffon had thought, but rather in the hundreds of millions of years.

Consider the ocean, for instance. It was 3.3 percent salt, and this salt, it appeared, was brought into the ocean at a trifling rate by the rainfall that scoured the continents, dissolved traces of material from the rocks and soil, and then carried those traces to the ocean.

* Ignorance has been waving the Bible ever since. Even today, there is nonsense called "scientific creationism" that is trying to foist itself on our schoolchildren—as I explained in the introduction to this book.

If one calculated the salt content of rivers, went on to calculate how much water (and therefore how much dissolved salt) the rivers delivered to the ocean each year, and assumed the oceans had started off as fresh water, it could be calculated that it would take about a billion years for the ocean to get as salty as it is now. That would mean the earth was at least a billion years old, or at least one eon old, if we define an eon as a billion years.

Some geologists would not accept the uniformitarian theory and yet could not deny the age of the earth. They suggested a series of catastrophes (*catastrophism*), each one shaping a new planet, with the last version of the planet being that of Genesis. William Buckland, an English geologist who was also in holy orders, held firmly to the Bible, but admitted there might have been millions of years during which a "pre-Adamite" earth had existed.

One of Buckland's pupils was Charles Lyell (1797–1875). He was a uniformitarian and his *The Principles of Geology,* published in three volumes from 1830 to 1833, destroyed catastrophism.

A long age for earth based on geological observations pleased the biologists, who, in the course of the first half of the nineteenth century, were coming to understand that there were forms of life that had lived and died on earth many, many years ago. Evolutionary notions were in the air and more and more biologists were beginning to risk theological thunder by supposing that life had not been formed in the twinkling of an eye by divine fiat but had very slowly evolved as a result of tiny, cumulative changes.

This reached its climax with the English naturalist Charles Darwin, a friend of Lyell's, who had been greatly influenced by his book. In 1859, Darwin published *On the Origin of Species by Means of Natural Selection,* which removed divine intelligence from the task of creating earth's present life forms and substituted random and slow-moving evolutionary change. For that, the one thing that was needed was time, and lots of it. A billion-year lifetime for the earth was none too long.

Against the combined force of geology and biology, however, were physics and astronomy. In the 1840s, the law of conservation of energy came to be accepted by physicists, with the German physicist Hermann Ludwig Ferdinand von Helmholtz (1821–94) its most influential advocate. Helmholtz undertook the problem of determining the source of the sun's energy.

Till then, no one had cared much. Earthly fires needed a constant feeding of fuel, but a heavenly fire was assumed to obey other rules, and it was an article of faith that the heavens were perfect and unchanging, at least until God was pleased to put an end to them.

But Helmholtz, following the law of conservation of energy, knew that the sun's energy poured out ceaselessly in all directions throughout history (and with only a vanishingly small fraction of it stopped by earth, while the rest tumbled wastefully past it) required a source—and a vast one.

By 1854, he had come to the conclusion that the only possible source (given the knowledge of the day) was contraction. All the substance of the vast sun was tumbling inward toward its center under its own gigantic gravitational pull, and the kinetic energy of that enormous fall was turned into radiant energy. This might sound as though the sun had not long to shine, but so enormous was it that a contraction of a mere ten-thousandth of its radius would supply it with a 2,000-year quantity of energy, at the rate it was radiating that energy.

Over the entire duration of civilized man, the necessary shrinkage of the sun required to have kept it shining would have been quite small, and certainly unnoticeable to the casual observer.

What, however, if vast ages of prehistory were included? The Scottish physicist William Thomson, later Lord Kelvin (1824–1907), calculated that if the sun had been radiating at its present rate for 25 million years, then to supply energy at its present rate during all that period it would have had to shrink from a radius of 150 million kilometers to its present radius of 1.4 million kilometers. And if the sun's radius were 150 million kilometers at some time in the past, its vast bulk would then have filled the earth's orbit, and the earth could only have been formed and cooled once the sun had shrunk to a considerably smaller size. By this reasoning, the earth could not be more than 25 million years old.

Kelvin tackled the problem from two other angles. He was aware that tidal friction was slowing earth's rotational speed and that, in the past, it had rotated more quickly than it does today. When it was hot enough to be molten, earth's rotation produced an equatorial bulge. Kelvin calculated at what speed the earth would have to rotate to produce an equatorial bulge of the size that actually exists. That turned out to be a rotational speed that the earth had about 100 million years ago, so that was about when it would have had to solidify.

Finally, if earth had been originally part of the sun, as astronomers at the time suspected, it would have begun its life at the temperature of the sun—say 4,500° C. How long would it have taken for earth exterior layers to cool down to their present pleasant temperature? Kelvin decided again that the probable answer was 100 million years, but toward the end of the century, he recalculated his figures in the light of newer knowledge of the physical properties of the earth's crust and decided it might be as little as 20 million years.

Through the second half of the nineteenth century, then, physics, the most respected and irrefutable of the sciences, gave earth a lifetime of not more than 0.1 eon and possibly as little as 0.02 eon, much to the discomfort of the geologists and biologists, who wanted and needed a much longer lifetime.

In 1896, however, radioactivity was discovered and physics was suddenly revolutionized. For one thing, a new source of energy was discovered. It was eventually called *nuclear energy* when, chiefly as a result of investigations of radioactive substances and their radiations, it was found that the atom had structure and that most of its mass was included in a very tiny nucleus at its center.

Each radioactive atom, as it broke down, gave off only a microscopic bit of energy, but all the atoms in the earth's crust as they broke down, bit by bit, gave off enough energy to keep the earth's crust at its present temperature indefinitely. That ruined the value of Kelvin's calculation. The earth could have been at solar temperature over a billion years ago, might have cooled off rapidly in 20 million years to its present temperature, then reached a point of stability and cooled off only very, very slowly thereafter.

The British physicist Ernest Rutherford (1871–1937), who was soon to discover the atomic nucleus, announced the role of radioactivity in this respect in 1904, with the aged Kelvin himself in the audience. Rutherford was very aware of Kelvin there and when he came to the crucial point and saw Kelvin's baleful eighty-year-old eye upon him, he quickly pointed out that Kelvin himself had said his conclusions were correct only if some unknown source of heat were not discovered. Kelvin's astonishing prediction was correct, said Rutherford; a hitherto unknown source of heat *had* been discovered. Thereupon Kelvin's face relaxed into a smile.

(Nevertheless, Kelvin was not flattered into accepting the new view. He died in 1907 and to the end refused to accept the newfangled notion of radioactivity.)

The other arguments for a short-lived earth also failed. It was quickly realized that nuclear energy offered a much more likely source for solar radiation than anything else did, and the whole notion of a Helmholtzian contraction of the sun was thrown out the window. If physicists relied on nuclear energy instead, the sun might well have been shining for an eon or more without any noticeable change in size. The actual details were worked out in the 1930s.

As for the earth's rotation and the size of the equatorial bulge, that assumed that once the earth solidified, its crust underwent no further major changes. Evidence accumulated, however, of the reverse and Kelvin's argument was abandoned. The discovery of plate tectonics in the 1950s and 1960s was the final crusher in this respect.

Rutherford, in his studies, showed (again in 1904) that a particular variety of radioactive atom broke down at a fixed rate. Any single atom might break down at an unpredicted moment, but a very large number of atoms of a particular variety, taken together, followed the rules worked out for what is called a *first-order reaction.*

In such a reaction, half the atoms would have broken down after a certain interval of time, say *x* years. Half of what remained would break down after an additional *x* years; half of what then remained after yet another *x* years, and so on. That period of *x* years, Rutherford called the *half-life.* Each variety of radioactive atom had its characteristic half-life, from very short to very long.

It turned out to be not a difficult task to determine half-lives. Uranium, the first element found to be radioactive, consists of two varieties of atoms, uranium-238 and uranium-235. The first has a half-life of 4.5 billion years, or 4.5 eons. The latter has a half-life of 0.7 eon. The element thorium consists of thorium-232 atoms exclusively, and this has a half-life of 13.9 eons.

Again in that same year of 1904, the American physicist Bertram Borden Boltwood (1870–1927) produced the final evidence that uranium-238, uranium-235, and thorium-232 are each the parent of a rather long series of radioactive descendants, all the members of which remained in uranium and thorium minerals in a delicate equilibrium if the minerals remain solid and undisturbed.

In 1905, Boltwood pointed out that lead was always found in uranium-containing minerals and that it might be the stable end product of the series. This turned out to be true. Uranium-238 initiates a radioactive series that ends in stable lead-206; uranium-235 ends in stable lead-207; and thorium-232 ends in stable lead-208.

Thinking about it further, Boltwood pointed out in 1907 that it might well be possible to determine the age of a rock—or at least the length of time during which it remained solid and undisturbed—in this way. By measuring the quantity of uranium and lead in a piece of rock and knowing the half-life of the uranium-238 atoms, we could determine how long it had taken that much lead to form from that much uranium.

There is, however, a catch. Lead doesn't exist only because it is formed from the breakdown of uranium and thorium. Lead exists in the earth's crust independently of radioactive breakdown. There is lead in rocks that contain no uranium or thorium to speak of and probably never did.

In that case, if we have a rock that contains both uranium and lead, it might be that all the lead formed from uranium and the rock has lain undisturbed for a very long time; or that most of the lead was there from the start and that very little was added through uranium breakdown, in which case the rock might be very young.

How can we tell what the true answer might be?

Let's consider lead, then. Lead is made up of four stable isotopes: lead-204, lead-206, lead-207, and lead-208. Of these, lead-206, lead-207, and lead-208, might have been there from the beginning, but some might also have been formed from the breakdown of uranium-238, uranium-235, and thorium-232 respectively.

Lead-204, however, is unique among the lead isotopes in not being formed from the breakdown of any naturally occurring radioactive atom whatever. Any lead-204 that exists in the earth's crust was present when the earth was first formed.

Suppose, next, we consider lead taken from sources where there has never been any uranium or thorium present, as far as we can tell. When this is so, the proportion of the various isotopes of lead is as follows:

lead-204 = 1.0
lead-206 = 15.7
lead-207 = 15.1
lead-208 = 34.9

If, in any rock that has both uranium and lead, you first ascertain the amount of lead-204 and multiply that by 15.7, you will determine the amount of lead-206 that was there from the beginning. Any amount of lead-206 present over and above that is there only because of the breakdown of uranium-238. Knowing the half-life of uranium-238 to be 4.5 eons, you can then tell how long it took that much extra lead-206 to form and, therefore, how long the rock has been in its solid state.

(Naturally, if the rock were at any time to become liquid, the uranium atoms and lead atoms would move about freely, undergo different chemical reactions, and separate. It is only while the rock is continuously solid that they are imprisoned and must remain together and in place.)

Any amount of lead-207 present to the extent of more than 15.1 times the amount of lead-204 is there because of the breakdown of uranium-235. Any amount of lead-208 present to the extent of more than 34.9 times the amount of lead-204 is there because of the breakdown of thorium-232.

Uranium-235 is present in uranium to a far lesser extent than uranium-238 is. Thorium-232, though about as common as uranium-238, has a half-life three times as long as that of uranium-238 and, therefore, breaks down with only a third the speed. The result is that uranium-235 and thorium-232 each contributes less lead than uranium-238 does, and it is the lead-206 content that is most useful in determining the age of rocks. Nevertheless, it is desirable that all three breakdowns give results that are in the same ballpark, as otherwise there is a strong suspicion that something is wrong.

On the whole, results were fragmentary. The earth is a geologically active planet. The actions of water, air, and life leave few sections of the crust untouched. Volcanic action and the shifting of crustal plates represent continuous large-scale changes. To find a piece of rock that has remained untouched and solid for a long, long time is difficult.

Nevertheless, such rocks were found and the results from the various lead isotopes did agree. Some rocks on earth have clearly remained untouched not merely for 1 eon, but for 2 and even for 3. A piece of granite from an African region formerly called Southern Rhodesia and now called Zimbabwe has been found to be 3.3 eons old.

An age of 3.3 eons for earth represents only a minimum. The earth's history has undoubtedly been more turbulent in its infancy than it is at present in its sedate middle age. The oldest rocks are either buried more deeply than we can reach, or else they simply don't exist anymore. Perhaps every scrap of solid material in earth's crust has at one time or another been melted, cooled, melted, cooled, several times during the early stages of earth's existence so that very little much older than 3 eons survives untouched—and that only by accident.

How much older than 3.3 eons earth might be cannot be determined, perhaps, by a study of earth itself.

Is there any way out?

Yes, there is. All current theories of the earth's origin assume it to have been formed along with the rest of the solar system in a single process. In other words, if we knew how old the moon was, or Mars, or the sun, we would know how old the earth was. What we are really searching for is the age of the solar system.

In general, all things being equal, the smaller the world, the more likely it is to be geologically dead, and the more likely it is that portions of it have remained solid from the earliest days of the solar system. This means that we'd more easily determine the age of the solar system, and of the earth, if we could analyze the surface rocks of the moon than we could by analyzing anything on earth.

Eventually, in 1969, we reached the moon, and obtained lunar rocks to analyze. Their ages, at least in the highlands, tend to run from 4 to 4.2 eons, older than anything on earth. Yet that, too, is a minimum. Earlier than 4 billion years ago (and no telling for how long a time before that if we had only lunar rocks to go by) the moon's crust was largely pulverized and destroyed and perhaps to a large extent fused by the collisions that placed all those craters and maria on the moon.

And there's a way out of that, too. In fact, we had something better than the moon, before we had the moon. There are the meteorites—bits of solid debris that may well have been circling the sun, formed and untouched even while the moon was shuddering under the final blasts of the matter that coalesced to form it.

Of the two major forms of meteorites, the iron meteorites do not contain significant quantities of uranium or thorium. The small amounts of lead they contain, judging from the lead-204 content, have been there from the start.

The stony meteorites, however, contain uranium and thorium in sufficient amounts to make an age estimation possible, and they prove to be about 4.5 eons old. This, by an odd coincidence (and it is nothing more) is just enough for half the original uranium-238 content to have broken down.

There are other methods of determining the age of the solar system and it isn't necessary to go into them. The important thing is that they all agree surprisingly well so that astronomers are confident that the solar system (and the earth) is 4.6 eons old—4.6 billion years.

That means that the oldest rock we have yet found on earth has been solid and untouched for 72 percent of earth's total history.

And yet that's just the solar system. The solar system is a speck of matter embedded in our galaxy, and it is in turn just one of many, many galaxies.

Was the solar system formed at the same time as our galaxy was, and the universe generally? Or is our sun and its train of planets a Johnny-come-lately, born into a universe already some eons old, or even countless eons old?

We'll take up that matter in chapter 13.

B ★ THE STARS

8 ★ THE RUNAWAY STAR

No one is as dangerous to an aging self-confident lecturer on the platform as a bright twelve-year-old in the audience. In the first place, at twelve years old the brightness of the brain is polished to a high sheen that has not yet been obscured by the light fog of healthy doubt. In the second place, a twelve-year-old monster of brightness is as yet unsubjected to any sense of decency or humanity. All he wants to do is to show off.

I know. I was once a bright twelve-year-old.

I was lecturing once on astronomy when a hand shot up in the audience. The owner of the hand was clearly a twelve-year-old with that sparkling eager look in his eye that I recognized at once. I would not have dreamed of recognizing him but his was the only hand in a sea of uninterested nonhands and I had to.

He said, in the invariable treble of a bright twelve-year-old, "Sir, which is the second closest star?"

I relaxed. I saw what his nefarious plan was. Everyone and his brother knew the closest star. It was Alpha Centauri. Very few, however, knew the second closest and the monster wanted to expose my ignorance. I smiled benignly for I knew I happened to be one of the very few who knew not only the name of the second closest star, but its distance, too.

I said, "It is Barnard's star, young man, and it is about six light-years from us."

Whereupon he looked puzzled and said, "That's funny. Then what's the closest star?"

I said, patiently, "That is Alpha Centauri, young man, which is four point three light-years from us and is actually a three-star system of—"

"But, sir," said the monster, springing his trap. "I thought that the sun was the closest star."

At once the audience woke up from its light doze in order to break into piercing cackles of laughter, something I helped along by standing on the platform with laughter of my own. (I have never cured myself of the bad habit of laughing at jokes at my own expense.)

I'm sure you would all be delighted to hear that I sought out the monster afterward and eradicated him from the face of the earth, but the truth is I did not. He is probably in graduate school now and is approaching the day when he will be on the platform facing a bright twelve-year-old and I hope he gets massacred.

What I will do for revenge is discuss Barnard's star and begin by asking who the devil Barnard was that he should own a star.

Edward Emerson Barnard was born in Nashville, Tennessee, on December 16, 1857, to an impoverished family. His father was already dead when baby Edward was born. He had time for exactly two months of formal schooling and at the age of almost nine he was working to help support himself and his family. He worked for seventeen years in a portrait studio, and that had its points, for it gave him an opportunity to learn the infant art of photography which was on its way to becoming the mainstay of astronomy. The telescope only aided the eye; the camera, to a large extent, replaced it.

Even while Barnard grew to be an expert photographer, he also developed an interest in astronomy and, while still a young man, discovered (that is, was the first to observe the approach of) several comets. As it turned out, Barnard had (it is believed) sharper eyes than any other astronomer on record. As an example, he once managed to detect a crater on the surface of Mars but did not report it officially because he thought he would be laughed at. Craters were indeed discovered on Mars in 1965 but not by the direct use of lesser human eyes than those of Barnard. They were photographed by the Mars probe Mariner 4 on a close approach to the planet.

Between 1883 and 1887, thanks to Barnard's growing reputation as an astronomer, he was appointed instructor at Vanderbilt University in Nashville. When he left (still without a degree) he joined the staff of the newly founded Lick Observatory at Mount Hamilton, California, and began his true professional career. In 1893, he received a doctor of science degree from Vanderbilt (for his work, not his earlier academic studies) and in 1895 became professor of astronomy at the University of Chicago. He worked at Yerkes Observatory after it was set up at Lake Geneva, Wisconsin, in 1905. He died on February 6, 1923.

In the fall of 1916, Barnard attended a meeting of the American Astronomical Society at Swarthmore College, near Philadelphia, and announced that he had noticed a shift in position of a particular star —a shift that meant it had a larger proper motion than any other star as far as was then known. That star has been called Barnard's star ever since. In view of its fast proper motion, it is sometimes called Barnard's runaway star.

The proper motion of a star is its change of position with respect to the various stars about itself and is usually expressed in seconds of arc per year, which can be abbreviated as ″/yr.

There are three factors that can contribute to the proper motion. One is the true relative motion of the star with respect to ourselves, the second is the fraction of that motion which is across our line of sight, and the third is the distance of the star.

As far as the first factor is concerned, all stars move relative to each other, and the relative speeds are, generally speaking, much the same, give or take a moderate amount. The true relative motion of the stars contributes to only a minor degree to any remarkable proper motion.

As for the second factor, most stars travel obliquely with respect to ourselves, neither directly toward us nor away from us, nor directly across our line of sight. The angle of motion also contributes to a minor degree only where any remarkable proper motion is concerned.

It is the third factor which is most crucial. For a given speed in a given direction relative to ourselves, the farther away a star, the smaller its apparent drift against the general background of stars; that is, the smaller its proper motion. In fact, for all except the nearest stars, the proper motion is too small to measure. Consequently the mere fact that a star has a proper motion at all tells an astronomer at once that it is a near neighbor to ourselves.

It is a good thing most stars have no measurable proper motion. It would be difficult to isolate individual proper motions otherwise, since the position of a particular star at a particular time can only be measured accurately relative to a neighbor star. If that neighbor star were itself visibly moving, conditions would grow complicated.

As it is, any star near enough to the solar system to have a proper motion need only have its position measured against the nearest convenient star, and we can almost always rely on that reference star being essentially motionless with respect to other stars and, therefore, a usable reference.

Barnard's star, once its position was measured with reference to nearby stars over a period of time, proved to be moving at the rate of 10.31″/yr. Not only does Barnard's star have a larger proper motion than any other star known at the time of its discovery, but no star with a larger proper motion has been discovered since to my knowledge.

Consequently, not only can we feel that it is a neighbor star, but it must be one of the closest there is, if not *the* closest.

Well, it isn't *the* closest. The closest star (always excepting our own sun, you rotten twelve-year-old kid) is Alpha Centauri, which is 4.27 light-years away. Barnard's star, however, is a good second, for it is only 5.86 light-years away.

To be sure 10.31″/yr is not much of a figure in itself. It is only amazingly large compared to the proper motion of other stars.

For instance, the width of the full moon, when it happens to be at perigee and is closest to the earth, is 33.50 minutes of arc (which can be written as 33.50′) in width. Since 1′ = 60″, the moon is then 2010″ in diameter and it takes Barnard's star 195 years to move the full width of the moon across the sky.

A degree of arc is symbolized ° and 1° = 60′, while there are 360° to a full circle. We can conclude that if Barnard's star were to move in the same direction at the same speed indefinitely, it would take 125,700 years to make a complete circuit of the sky.

Why does Barnard's star have a larger proper motion than Alpha Centauri even though Barnard's star is the more distant? Alpha Centauri's proper motion is 3.68″/yr, or only about a third that of Barnard's star.

At the distance of Barnard's star, a proper motion of 10.31″/yr corresponds to a transverse motion (one across the line of sight) of about 90 kilometers per second (km/sec). The radial motion (directly toward or away from us) of Barnard's star can be determined from the shift of the dark lines of its spectrum as compared with light from a stationary source. The shift in the case of Barnard's star is the equivalent of a speed of 108 km/sec. Since the star exhibits a shift in the direction of the violet end of the spectrum, it is approaching us at that speed.

Combining the two speeds, we find that the "space velocity" of Barnard's star is 141 km/sec toward us at an angle of 50° from the line of direct approach. Subjecting Alpha Centauri to the same analysis, its space velocity is 34 km/sec toward us at an angle of 47.5° from the line of direct approach.

Both stars are approaching us obliquely at nearly the same angle so the angle of motion is not a factor. However, Barnard's star's space velocity is 4.14 times that of Alpha Centauri and that, in *this* case, is the crucial factor. The greater distance of Barnard's star reduces the effectiveness of its greater space velocity so that its proper motion is only 2.8 times as great as that of Alpha Centauri.

Barnard's star's space velocity is by no means a record even among those nearer stars where the speed can be determined with some hope of precision. Consider Kapteyn's star, for instance, which is named for the Dutch astronomer Jacobus Cornelius Kapteyn (1851–1922). It has the second largest proper motion known, 8.79″/yr, nearly seven-eighths that of Barnard's star. Kapteyn's star, however, is 13.0 light-years from us, so that it is two and a quarter times as far away from us as Barnard's star is.

Kapteyn's star is receding from us at an angle of only 34.5° from the line of direct recession, so a distinctly smaller fraction of its motion shows up across the line of sight and can be measured as proper motion. The space velocity of Kapteyn's star, however, is 293.5 km/sec, so that it moves twice as fast as Barnard's star does.

If the motions of Alpha Centauri, of Barnard's star, and of Kapteyn's star were, all three, directly across the line of sight, then the proper motions would be, respectively, 5.44″/yr, 16.15″/yr, and 15.54″/yr. Barnard's star would barely hold its primacy over Kapteyn's star in that case.

Of course, Barnard's star is approaching us and Kapteyn's star is receding from us so that, as the centuries pass, Barnard's star will increase its proper motion because it will be closer to us and also because, as it approaches, a larger and larger fraction of its motion will be across our line of sight. Kapteyn's star will, on the other hand, decrease its proper motion for the opposite reasons. (In both cases, I am assuming that the speeds of each with respect to the solar system will not change in the reasonably short run.)

Barnard's star will make its nearest approach to us 9,800 years from now, at which time it will be 3.85 light-years away. This is only three-fifths its present distance. At that point of closest approach, Barnard's star will be moving directly across our line of sight.

Its proper motion at that time will be 26.4″/yr, or two and a half times what it is now. If it is a runaway star now, what will it be then? It will be moving at a rate that will carry it the width of the full moon at perigee in a mere 76 years.

Will Barnard's star be the closest star to us at that time?

The only competitor it can possibly have will be Alpha Centauri. Barnard's star will be closer, at its closest, than Alpha Centauri is *now,* but Alpha Centauri is also approaching us at a slant. When Barnard's star is at its closest, 9,800 years from now, Alpha Centauri will be about 3.92 light-years from us, a third of a light-year closer than it is now, but it will then be not quite as close as Barnard's star will be.

If my rough calculations are correct, Barnard's star, for a comparatively brief period of time centering about a period ten millennia from now, will be the closest star in the sky (always excepting the sun, you rotten twelve-year-old kid).

But at that time, Barnard's star will be passing us transversely and will begin to recede and will continue to do so for a very prolonged period. At that time, and for thousands of years afterward, however, Alpha Centauri will continue approaching us at a slant. It will be at its point of closest approach about 38,000 years from now, when it will be moving transversely past us at a distance of 2.90 light-years, only two-thirds its present distance. Then it, too, will begin to recede.

An interesting point can be made if we measure these minimum distances in parsecs (which astronomers prefer to light-years). One parsec is equal to 3.26 light-years. That means that Barnard's star will approach from its present distance of 1.81 parsecs to a minimum distance of 1.18 parsecs 9,800 years from now. Alpha Centauri, on the other hand, will approach from its present distance of 1.31 parsecs to a minimum distance of 0.89 parsecs 38,000 years from now.

For a brief period (astronomically speaking) in its path through the galaxy, Alpha Centauri will be less than a parsec away from the solar system.

This is not unusual in the long run. Paul R. Weissman of the Jet Propulsion Laboratory estimates that a star passes within a parsec of the solar system every 200,000 years on the average. This means that 23,000 stars have done so in the 4.6-billion-year lifetime of the sun, and perhaps 30,000 more will do so before the sun ends the present stage of its life cycle and becomes a red giant.

Still, knowing that lots of stars will do so at some time or other is not the same as knowing that some particular star will do it at some particular time.

So I have a question to ask of any of my Gentle Readers who have better data available to them than I have and who know more about celestial mechanics than I do. We know that Alpha Centauri is on its way to passing within a parsec of ourselves, but will any other star follow it that we know of? Is there a particular star (or stars) in the sky whose space velocity is known and which is aimed at us with such accuracy that it will at some known time pass us within a parsec's distance? If so, which star (or stars) is it and when will it happen? Anything I am told in this respect, I will share with all my Gentle Readers.

The proper motions of stars were first noted in 1718 by the English astronomer Edmund Halley (1656–1742). He noted that Sirius, Procyon, and Arcturus were well removed from the positions in which ancient observation had placed them. As it happens, these stars have rather small proper motions; that of Sirius is 1.32″/yr and that of Procyon is 1.25″/yr, but Halley had the advantage of two thousand years of observation.

Once Halley had made his observation, proper motions were searched for in connection with other stars. How did it come about, then, that the proper motion of Barnard's star, the fastest moving of them all, was not detected till 1916, two hundred years after the discovery of the phenomenon?

The answer is, of course, that Barnard's star did not have the advantage of thousands of years of observation because it is not a bright star. In fact, it is not visible to the unaided eye, so that it could not have been seen till after the invention of the telescope in 1608. The magnitude of Barnard's star (that is, its apparent brightness on a logarithmic scale) is 9.5, and the dimmest star that can be seen by good eyes on a dark, clear moonless night may be 6.5

in magnitude. (The smaller the number representing the magnitude, the brighter the object.) This means that Barnard's star is only one-sixteenth as bright as the dimmest star that can be made out by the unaided eye.

Even when the use of telescopes made it possible to see Barnard's star, it was only one of about 130,000 similar faint stars and there was no reason to observe *it* painstakingly rather than any of the others. That *it,* virtually alone among them, had a huge proper motion could be noted only by accident. Someone, comparing two views (or photographs) of a star field that happened to contain Barnard's star, would have had to notice that one of the powdering of stars was out of place, look more closely, say "Hey, that's funny," and begin to make other views (or look for other plates of the same star field taken at still other times.)

It happened to be Barnard who did it, and he immortalized his name in consequence.

Barnard's star is a small star, with not more than a fifth the mass of our sun and therefore only two hundred times as massive as the planet Jupiter. It is not very much more massive than the minimum required to produce enough temperature and pressure at its center to ignite the hydrogen fusion reaction. The fusion that does take place does so at a comparatively low rate so that the surface temperature of Barnard's star is only 2,800° C., or just half the surface temperature of the sun.

At that temperature, Barnard's star gleams with only a dull red light. It belongs to spectral class M5 and is a red dwarf.

If our sun were replaced by Barnard's star, it would be a red circle in the sky only about a quarter the diameter of the sun, and in apparent area it would be about a sixteenth that of the sun.

The total amount of light we would get from Barnard's star would then be $1/_{2750}$ that which we now get from the sun. The warmth we would receive from Barnard's star would be an equally small fraction of that we get from our sun so that earth would be an eternally frozen wasteland if Barnard's star became our luminary. In order to get as little heat from our sun, earth would have to revolve about it at a distance of something like 7.8 billion kilometers or just a bit farther out than Pluto at its farthest.

If earth were to remain as far from Barnard's star as it is from the sun, it would be gripped in a feebler gravitational field and would make its circuit about Barnard's star in something like 850 days or rather over two and a quarter years.

Of course, we needn't look down at Barnard's star with too much contempt. It is not at all the dimmest star there is. For instance, the feeble distant companion of the Alpha Centauri binary star system, which is known as Proxima Centauri is also a red dwarf, but one that is distinctly dimmer than Barnard's star.

Barnard's star is nearly seven times as luminous as Proxima Centauri is.

Still, Barnard's star is approaching us and its apparent brightness is increasing, which means its magnitude is decreasing. What about the time, 9,800 years from now, when Barnard's star makes its closest approach? How bright will it be then? Will it then be visible to the unaided eye?

At its closest approach, Barnard's star will be 0.9 magnitude lower than it is now. Its magnitude will be 8.6. It will be two and a quarter times as bright as it is now, but it will still be only one-seventh as bright as necessary to be just made out by unaided eyesight under the most favorable conditions from earth's surface. Barnard's star would have to be within 1.5 light-years (half a parsec) from earth to be just barely made out as a dim star to the unaided eye.

Nevertheless, dim or not, Barnard's star's proper motion is spectacular now and will be more spectacular at its closest approach. Imagine astronomers photographing the region of the sky in which Barnard's star is located on every New Year's Day. In seventy-six years, it would move the width of the full moon while the stars about it would not move any perceptible amount. If the seventy-six frames were superimposed and flipped past a projection machine at a moderate speed, Barnard's star would be seen creeping across the star field. It would give people the visible feeling of the motion of that nearby runaway star.

Since we've considered the increasing brightness of Barnard's star what about the increasing brightness of Alpha Centauri?

Right now, Alpha Centauri has a visual magnitude of —0.27 (which, of course, takes into account the combined light of both members of the binary system, since these cannot be separated into two different light sources by the unaided eye). This is very bright for a star, which is not surprising, since Alpha Centauri is so much closer to us than other stars are.

Yet Alpha Centauri is only the third brightest star in the sky. It is outdone by two other stars that are each farther than Alpha Centauri is, but that are so much more luminous intrinsically that even at their greater distances they shine more brightly than Alpha Centauri does.

The second brightest star is Canopus, with a magnitude of —0.72, while the brightest of all is Sirius, with a magnitude of —1.42.

But here again, we must consider that Alpha Centauri is approaching earth and is slowly increasing in brightness. How bright will it be 38,000 years from now when it reaches its minimum distance of 2.90 light-years from the solar system?

At that time, it will be shining in the sky with 2.17 times the brilliance it now possesses. That means that it will have a magnitude of —1.11 and it will then be brighter than Canopus by a good bit.

Canopus, which is about 195 light-years away, nearly fifty times the distance of Alpha Centauri (and yet so luminous as to outshine Alpha Centauri in our skies today), is receding from us. In a mere 38,000 years, however, it will have receded only about 2.6 light-years and the increase of its distance by merely 1.3 percent will raise its magnitude from —0.72 to —0.71, an insignificant dimming.

Nevertheless, even at a magnitude of —1.11, Alpha Centauri will remain dimmer than Sirius is today. And it should be remembered that Sirius is also approaching us, albeit very slowly. It has a space velocity of 18 km/sec at an angle of 63° to the line connecting us. It will, after 38,000 years, be just about 1 light-year closer than it is today—say 7.63 light-years away instead of 8.63. It will then be shining about 1.28 times as brightly as it is now and its magnitude will be —1.69.

Alpha Centauri will never be the brightest star in the sky, therefore. At its brightest it will be shining only 60 percent as brightly as Sirius will be shining at that time. (At that, it's an improvement; Alpha Centauri shines only a third as brightly as Sirius today.)

After Alpha Centauri's closest approach 38,000 years from now, however, it will begin to recede and fade while Sirius will continue to approach. It will take about 60,000 years for Sirius to skid past us at its point of closest approach when it will be 7.15 light-years away from us. Its magnitude will then be —2.27 and it will be shining just a bit over twice as brightly as it is now.

There will be many stars that will skim past earth more closely than Sirius will. Alpha Centauri and Barnard's star are two of them, for instance. However, most of the skimmers will be dim stars, since dim stars are far more numerous than bright ones. Not more than one star in a thousand is brighter, intrinsically, than Sirius is, so that it will shine more brightly in the sky when it approaches us as closely as Sirius will, or even somewhat less closely.

So here is a second question for you experts. Is there any star in the sky, of whatever size, that we know is going to pass us in such a way as, at its point of closest approach, to shine with a magnitude of less than —2.27 so as to be brighter than Sirius at *its* closest approach? If this is so, what is the name of the star (or stars) and when will it happen and how bright will the star (or stars) get? Again, any information I receive will be shared with the Gentle Readers.*

However, we have not yet finished with Barnard's star. Its most interesting aspect remains to be considered—in the next chapter.

9★ THE DANCE OF THE STARS

My wife, Janet, and I were hastening to catch a train the other morning. The trouble was that it *was* morning, the morning rush hour in fact, with winter slushiness underfoot so that no one wanted to walk even short distances. That meant there would be no taxis available.

Since the train wouldn't wait for us, and we could scarcely walk a mile and a half through the slush with our bags, we decided to take the bus.

Came the bus! People crowded into it until no more would fit—and there we were, still outside. We then noted that people were also getting into the entrance in the rear, which was not an entrance at all but an exit. We were in no mood to quibble. We raced over and we were the last two in. I just barely fit.

There is no one at the rear door to accept fares, however, so the people who get in there do not pay fares at all—behavior I consider unsocial and uncivilized.

* Daniel U. Thibault of the University of British Columbia wrote to me to tell that a star known as Ross 248 will approach a little closer than Alpha Centauri or Barnard's star will ever manage. Ross 248 is another dwarf star, however, and he can find no star that will outshine Sirius in the reasonably near future. Gordon Palameta writes to tell me, however, that Canopus, about 1,400,000 years ago, was only 17 light-years away and shone as brightly as Venus at its brightest.

As we were approaching Penn Station, we found the bus's population density had thinned to the point where movement was possible. Janet and I therefore made our way forward, and Janet said, "Driver, we entered the bus at the rear and did not pay, so here are our fares." She tossed in her coins and I tossed in mine.

The driver glowered at us and growled, "Do you know you could be arrested for getting in at the rear?"

I stifled the impulse to ask, "Arrested by whom?" since the last time any policeman has been seen on the streets of New York was sometime in 1967, and I merely sighed. We were the only ones who, having gotten on in the rear, were conscientious enough to pay our fares—so *we* got the threat.

But that's the way the universe works, which brings us back (at least eventually it will) to the subject of the previous chapter—Barnard's star.

If we ignore planetary perturbations, which are very small, we might say that the earth travels about the sun in a smooth and geometrically neat ellipse. And if we say so, we would be wrong.

It is not the earth itself that marks out a smooth ellipse about the sun; it is the center of gravity of the earth-moon system.

The center of gravity of the earth-moon system is always on the imaginary line connecting the center of the moon to the center of the earth. Since the earth has 81.3 times the mass of the moon, the center of gravity is 81.3 times as close to the earth's center as to the moon's.

This means that, on the average, the center of gravity of the earth-moon system is located 4,728.2 kilometers (2,938 miles) from the center of the earth. When the moon is at perigee and is closest to the earth, the center of gravity is correspondingly closer to the center of the earth, and when the moon is at apogee, and farthest, the center of gravity is correspondingly farther from the center of the earth. The difference is not great, however, only a matter of some 600 kilometers (370 miles) at most.

The moon travels about the earth in an elliptical orbit with the center of gravity of the earth-moon system at one focus of the ellipse. This is almost the same as saying that the earth is at one focus of the ellipse, for the center of gravity of the system is so close to the earth's center that it is actually inside the earth. Its depth, on the average, is 1,649 kilometers (1,025 miles) below the surface of the earth.

The moon also revolves around the sun, accompanying the earth, but it obviously does so in an orbit that is not a smooth ellipse, for it is sometimes on the side of the earth away from the sun and sometimes on the side toward the sun. The difference in the moon's distance from the sun, depending on which side of the earth it's on, is 766,000 kilometers (476,000 miles)—allowing for the fact that its orbit about earth is slightly tipped to its orbit about the sun.

This difference isn't much compared to the total distance of the earth-moon system from the sun. The difference is only about half a percent of the total distance in fact, so that if we were to mark out the moon's orbit about the sun to scale, it would look like a smooth curve, and one that was nearly circular to boot.

Suppose, however, we looked closely and examined the orbit as drawn to scale under a strong magnifying glass and made precise measurements of the distance of different parts of the orbit from the sun's center. We would then find a series of very shallow waves in the moon's orbit about the sun, a little over twelve of them in the course of the complete orbit.

Suppose we were observing the earth-moon system's travels about the sun from afar and, for some reason, could not see the earth but could observe only the moon. From the manner in which the moon deviates from the smooth ellipse, the distance to which it recedes from it and the time it takes to complete the wave, it would be possible to infer the existence of the unseen earth, to calculate the distance between the moon and the earth, and, if the moon's mass could be worked out, to deduce the mass of the earth.

If it were only the earth that was seen and not the moon, the data on the unseen moon could be worked out similarly, but with much more difficulty. The earth swings away from the ideal orbit with a wavelength precisely that of the moon, but an amplitude only 1/81.3 times as great. Therefore, the earth's movement must be plotted with much greater precision.

Of course, this is an artificial problem, since if we can see the earth we are virtually sure to be able to see the moon, too, and vice versa.

Even if the earth-moon pair were situated so far from us that the earth could only be seen through a good telescope, the moon would also be seen. We know this is so because Pluto is smaller than earth and its satellite, Charon, is smaller than the moon, yet though both are at a great distance from earth, both can be seen. The only reason it took nearly half a century, after Pluto was first seen, to spot

Charon as well is that the two are so close together that, at their vast distance, they melt almost into a single point of light. If Charon were as proportionately far from Pluto as the moon is from the earth, Charon would have been discovered immediately after the discovery of Pluto.

As it is, as soon as Charon was made out, then from its distance from Pluto and its period of revolution about Pluto, the mass of each world could be calculated, even though till then that of Pluto had been a puzzle.

There are cases, however, where, of a pair of bodies circling a common center of gravity, one is easily seen and the other cannot be seen at all. This would be true if one body were comparatively large while the other were very tiny. Better yet would be the case where one is enormously bright and the other comparatively dim and comparatively close to the bright companion. In that case, the minor object would be difficult to see not only because of its intrinsic faintness but also because it would be drowned out in the inordinate brilliance of the other object.

Suppose that, instead of considering the earth-moon pair, we were to consider the earth-sun pair.

If the sun-earth pair were so far away that the sun could only be seen in a good telescope there would be no hope, under the conditions of observation now available to us, of seeing earth at all. Earth would be too dim and it would be so close to the sun in appearance that it would be totally masked.

Nevertheless, might we not tell that the earth was there, even though we couldn't see it, simply by observing the sun's motion?

The sun moves about the center of the galaxy in a period of about 250 million years, and if it were alone and unaccompanied, it would do so in a smooth ellipse (barring the effects of gravitational perturbation produced by other stars, which we can assume will be small enough to neglect).

It is not the sun, however, that moves in that ellipse but the center of gravity of the solar system. If the solar system consisted only of the sun and the earth, then the center of gravity of the earth-sun system would be on the line connecting the centers of the earth and sun. Since the sun is 324,000 times as massive as the earth, it would be 324,000 times as close to the sun's center as to the earth's.

This means that the center of gravity of the sun-earth system would be 462 kilometers (287 miles) from the center of the sun, in the direction of earth. Therefore, as the sun progresses in its journey about the galactic center, it wobbles slowly from side to side with a period (of course) of one year.

The wobble isn't much, for it is only about one fifteen-hundredth the sun's radius, and detecting it would pose a pretty problem indeed for anyone observing the sun from, say, some planet circling Alpha Centauri.

But then, in observing the sun, we are not really observing the earth-sun system. There are other planets circling the sun and each one has a planet-sun center of gravity of its own. On the whole, we can say that the more massive a planet is and the more distant it is from the sun, the greater the displacement of the planet-sun center of gravity from the sun's center.

As it happens, the four gas giants, Jupiter, Saturn, Uranus, and Neptune are all considerably more distant from the sun than earth is and each is considerably more massive than earth is. Any irregularity imposed upon the sun's motion by earth's existence would be insignificant in comparison to the much greater irregularities imposed on it by the gas giants. And if earth can be ignored, so, certainly, can the vast number of bodies in the solar system that are even less massive than earth. Even distant Pluto and Charon, whose vast separation from the sun might be thought to impose a considerable displacement of the center of gravity, have so little mass that their effect on the sun's motion is less than that of earth.

Let's consider, then, the center of gravity of the systems that involve the gas giants only. Knowing the mass of each planet and its distance from the sun, it turns out that:

Planet	*Distance of Center of Gravity from Sun's Center*	
	Kilometers	*Miles*
Jupiter	763,800	475,000
Saturn	419,000	260,000
Uranus	129,000	80,000
Neptune	239,000	149,000

As you see, the greatest displacement of the sun is produced by Jupiter, thanks to its huge mass. The other three planets are considerably farther from the sun than Jupiter is, but they are also considerably less massive, and it is the latter effect that predominates in this case.

The center of gravity of the Jupiter-sun system lies outside the globe of the sun, since the sun has a radius of only 696,000 kilometers (432,000 miles). The center of gravity lies 67,800 kilometers (43,000 miles) above the sun's surface, in the direction of Jupiter.

As the sun moves in its orbit about the galactic center, then, it swings this way and then that, with an amplitude greater than its own radius and with a period of about 12 years. (Even if the sun were motionless, for that matter, it would still move from side to side and its center would make a tiny ellipse in the sky, thanks to Jupiter's motion about it.)

If we were viewing the sun from a planet circling Alpha Centauri, with instruments equal to the best we now have on earth, we might just barely be able to make out this dance of the sun. Its exact nature, whether it was symmetrical or distorted, whether it involved a change in rate of motion as well as a change in position, or whether rate of motion entirely replaced change in position, would depend on our angle of view.

In theory, whatever the position from which the sun was being observed (provided it was no farther off than Alpha Centauri) and whatever the rate of motion of the sun relative to ourselves, we could detect Jupiter and tell something about its properties if our measurements were delicate and precise enough.

Indeed, as we watched the sun and its motion very precisely, we would find that it would sometimes deviate more than usual, and sometimes less than usual, depending on whether Saturn was on the same side of the sun as Jupiter was, or on the opposite side. The position of the center of gravity of the solar system would depend not just on Jupiter but on the position of every object in it—and, overwhelmingly, on the relative positions of the four gas giants.

An observer from the Alpha Centauri system, if he were able to make *very* precise and prolonged observations might be able to detect from the sinuosity and changing speeds of the sun's motion the existence of four planets of particular masses and particular distances from the sun. Given enough time and enough precision, he might, in theory, detect even smaller bodies of the solar system.

(And, before I forget, just how does an observer note the direction and speed of motion of a particular star? By measuring its distance from some apparently nearby star that is actually so distant that its own motions, even over prolonged periods of time, are too tiny to be detectable and that therefore can act as a stationary reference point.)

It works the other way around, too. If, from our solar system, we observe the stars, we could, in theory, detect their wavering dances and tell whether they possess planets. We could tell how many, how massive, and how distant from the star those planets were, if we could observe the dances delicately enough.

However, the farther a star is, then the smaller the apparent dance produced by an orbiting planet. At even moderate distances (for stars) the apparent irregularities in their motions become so tiny that there is no practical hope whatever of direct detection of planets through them. We must therefore confine ourselves to the nearest stars—our immediate neighbors.

Even in the case of the nearest stars, we can only expect to detect giant planets—like Jupiter, or larger. Earth-sized planets would produce but an indetectable wobble in even the nearest star.

For that matter, even Jupiter won't do the job unless the star is less massive than the sun. (What counts is not how massive an orbiting object is in absolute terms, but how massive it is in comparison to the object it circles. The moon produces a considerable wobble in earth, but if it were circling Jupiter instead at the same distance, Jupiter's wobble would be inconsiderable.)

This has actually been put into practice. The star Sirius is 2.5 times the mass of the sun, yet it dances to an extent that is easily detectable; so easily detectable that it was detected a century and a half ago. But then, Sirius dances to the tune of a white dwarf star, too dim and too close to Sirius to be easily detectable, but with the mass of a thousand Jupiters. That's not the same thing as detecting a planet.

Well, then, can we detect, at stellar distances, the much tinier dance that would reflect the presence of a planet and not merely another star? We might! In fact, we may have done so!

There are about fifteen stars close enough to us and small enough to exhibit just barely visible irregularities in position if they had planets circling them that were like Jupiter (or more) in size and in distance from themselves.

The first case of this sort involved the star 61 Cygni (the sixty-first star in the constellation Cygnus (the Swan), so numbered according to a system invented by John Flamsteed (1646–1719), the first astronomer royal of England).

As it happens, 61 Cygni is close to us, as could be inferred from the fact that until Barnard's star was discovered, 61 Cygni had the fastest known proper motion. In fact, F. W. Bessel (1784–1846), in his attempt to determine the distance of a star, chose 61 Cygni as his victim for precisely that reason. He managed to measure the star's parallax and announced the result in 1838 so that 61 Cygni has the distinction of being the first star to have had its distance determined.

It is actually 11.1 light-years from us, which amounts to 105 trillion kilometers (65 trillion miles). That makes it, of all stars visible to the unaided eye, the fourth closest to ourselves.

Actually, 61 Cygni is a binary star; two stars circling about a common center of gravity.

Each of the two stars is smaller than the sun. The larger of the two, 61 Cygni A, has a diameter only about seven-tenths that of the sun. The diameter of 61 Cygni A is about 965,000 kilometers (600,000 miles) while that of 61 Cygni B, the smaller of the pair, is about 900,000 kilometers (560,000 miles). The two stars, combined, have about two-thirds the mass of our sun.

The two stars of the 61 Cygni binary system are separated by an average distance of about 12.4 billion kilometers (7.7 billion miles), or a little more than twice the average distance between our sun and Pluto, and they circle each other about their center of gravity once in 720 years.

Either or both stars could have a planetary system that would not suffer undue interference from the other star, though the systems would probably have to be somewhat less extensive than the sun's, which lacks any companion star whatever.

If the planet earth were circling one of the 61 Cygni stars at the same distance it now circles the sun, that star would appear in the sky as a red-orange object distinctly smaller than the sun (which would mean that the earth would be frozen into a permanent ice age, of course). The other 61 Cygni star would be dimly visible, if it happened to be shining by night, as a point of light. It would be a bright starlike object, showing no visible disk.

In 1943, the Dutch-American astronomer Peter Van de Kamp (1901–) found an irregularity in the movement of the 61 Cygni stars about each other. From this, he deduced the presence of a third object in the system which he called 61 Cygni C and which was, of course, smaller than either star.

If 61 Cygni C were responsible for the irregularity, it would have to have a mass eight times that of Jupiter. It would be just too small to set up nuclear fusion at its core and shine by its own light, so it meets the usual definition of *planet*. This means that 61 Cygni C is the first extraterrestrial planet to be discovered.

Two Soviet astronomers have been studying the 61 Cygni system carefully in recent years, and, combining their observations with the earlier ones of Van de Kamp, have reported that the irregularity was itself irregular. In April 1977, they suggested that 61 Cygni A might have two planets, one with six times the mass of Jupiter and the other with twelve times the mass, while 61 Cygni B might have a planet with seven times the mass of Jupiter.

If so, 61 Cygni is not merely a binary star, but a binary planetary system. Undoubtedly, if each has one or two large planets, each could, and should, have a whole train of smaller planets, satellites, asteroids, and comets—all too small to leave detectable marks on the irregularity.

Nor is 61 Cygni the only star to display the presence of planets. Some half a dozen others seem to show the presence of super-Jupiters.

This is important. We know that stars are very numerous by direct observation, but we have no similar knowledge that planetary systems are. If planetary systems are very rare (as is possible) then there is no hope that other intelligent beings and other civilizations are to be found anywhere near ourselves. If planetary systems result only from some extremely unusual process in star formation so that our own solar system is one of only a handful in the entire universe—then we may be alone!

On the other hand, if planetary systems are common, and if they routinely accompany all but the most unusual stars, then there is a good chance that other civilizations exist, perhaps a very good chance. In fact, as some astronomers argue, other civilizations by the millions are inevitable.

From current theories of the origin of the solar system, it would seem that the latter alternative is more likely to be true; that virtually every star has a planetary system and that, therefore, civilizations may be common in the universe. Still, it would be nice if we did not have to depend solely on theory; if we had *some* observational evidence.

If in fact, then, of the very few stars that are close enough to show irregularities, half a dozen do, we must conclude that planetary systems are very common and possibly almost universal. If that were not so, the stringent requirements for detecting such planets would simply not be met.

And now back to Barnard's star—

Of the stars considerably smaller than the sun, Barnard's star is the second nearest to ourselves. Only Proxima Centauri is closer.

Barnard's star, moreover is a single star, so there isn't the possibly confusing fact of a second star near itself. What's more, its rapid proper motion should stretch out the wave of irregularity and make it perhaps the more noticeable.

And it *has* been noticed. Van de Kamp has found irregularities in its motion that are larger than those of any other star, and he interprets them in such a way as to show the presence of two planets circling the star.

Of these the one closer to the star, which we can call Barnard's star B, has about 1.1 times the mass of Jupiter; while the other, Barnard's star C, which is farther away, has about 0.8 times the mass of Jupiter. These planets are the least massive of any of those that have been reported circling other stars. In fact, the second planet is the only one yet reported that seems to be smaller than Jupiter.

Barnard's star B and C do not seem to be very different from the sun's Jupiter and Saturn. In fact, Barnard's star B circles the star itself (which we should call Barnard's star A) in twelve years, while Barnard's star C circles it in twenty-four years—as compared with twelve and twenty-nine years for our own Jupiter and Saturn.

All this is very exciting, except that here we come up to something as disappointing and deflating as that nasty bus driver Janet and I encountered.

All of this extrasolar planetary data depends on tiny deviations from the expected positions of stars that are just on the borderline of what can be detected.

In recent years, astronomers, observing Barnard's star very meticulously and with a variety of telescopes, have grown dubious over the reported irregularities. These *might* show the presence of planets, but they might equally well be the result of inevitable uncertainties of observation.

And if the results with reference to Barnard's star are made to seem dubious, then the irregularities found in other stars, all of which are even smaller than those of Barnard's star, are surely of uncertain significance.

This would mean that we can't have the confidence of having actually *observed* events that demonstrate the existence of planets. We can't have the security of feeling that planetary systems are very common and, perhaps, virtually universal. We can't have the glorious expectation that there may be other civilizations not too far away that someday we may be able to establish communication with.

Does that mean we have to give up?

Of course not. The *principle* remains untouched. If there are planets circling stars, then that will show up as an irregularity in the star's motion, and this irregularity will be greater, the smaller the star, the larger the planet and the greater the star-planet separation. All that is undeniably so. The only trouble is that even under the most favorable circumstances, the irregularities are too small to be measured with confidence.

In that case, we must change conditions of observation to make small irregularities detectable with greater precision. An obvious change for the better would be to place a large telescope on the moon, or in orbit. If we could observe the nearby stars without the interference of an absorbing, refracting, temperature-quivering atmosphere, that alone would remove an important source of fuzziness and error.

Then, too, once out in space, the full range of electromagnetic radiation can be observed and it may be possible to reduce the contrast between the brightness of a star and the dimness of its nearby planetary companion by choosing some proper wavelength, by the ingenious use of computers, and by other devices, so that we might actually *see* the planet.

If then we could unmistakably detect planets circling half a dozen of the nearest stars, we would be back to the position of universal planetary system and millions of civilizations. And if no planets turned up, we would have to accept that, too, and console ourselves with the thought that this might help us develop more accurate theories of star formation and a better understanding of the universe.

Either way, this alone would make worthwhile any expenditure of effort or money likely to be involved in putting a large telescope out into space.

C ★ THE UNIVERSE

10★ GETTING DOWN TO BASICS

When you're signing autographs, there's no use being grumpy about it. People are flattering you with their interest in you and they deserve at least the reward of friendliness and banter, so I try to provide it.

With experience, I've worked up a lot of banter items, the best of which are, of course, those which elicit smiles and are unanswerable so that we can move on. Occasionally, of course, the unanswerables are answered.

For instance, suppose a good-looking woman in her late youth hands me a book to sign and says (as she often does), "My son is simply mad about you, Dr. Asimov. He reads every book of yours he can possibly get hold of."

In that case, as I sign, I am quite likely to say, with a winning smile on my frank and ingenuous countenance, "How delightful! Imagine how pleased and honored he would feel if you and I were to have an affair."

There is bound to be a giggle and, since it is an unanswerable remark, it's on to the next.

Except for once, when, after I had suavely delivered myself of my pleasantly outrageous remark, the woman before me stood her ground and said, "Dr. Asimov, if we were to have an affair, *I* would be pleased and honored."

And with my remark neatly topped, all I could do was stare at her in pink-cheeked silence.

Serves me right, of course.

My only consolation is that in the game of science, which is the occupation of my more serious moments, having one's best answers proven insufficient is the common state of affairs. Consider the matter of getting down to basics, for instance.

Step One — The Greek Elements

The first person in our western tradition of rational inquiry who concerned himself with the basic composition of the universe was the Greek philosopher. Thales (624–546 B.C.). His answer was "Water."

Other philosophers had other suggestions and Aristotle (384–322 B.C.) listed four basic substances—earth, water, air, fire—for the world about us; and a fifth, ether, for the heavenly objects.

Each of these was eventually named *elementum* in Latin (*element* in English), a word which, oddly enough, is of unknown origin.

Step Two — The Chemical Elements

The doctrine of the five Greek elements lasted two thousand years. Then, in 1661, the English chemist Robert Boyle (1627–91) published *The Sceptical Chemist.*

Science had by then become experimental and inductive, rather than introspective and deductive as among the Greeks, and Boyle suggested that an element had to be tested in the chemical laboratory. If it could not be broken down into still simpler substances, then—and only then—it could be so labeled.

By this criterion, there were twelve elements known in Boyle's time: gold, silver, copper, tin, iron, lead, mercury, carbon, sulfur, arsenic, antimony, and phosphorus.

The number continued to grow. The French chemist Antoine Laurent Lavoisier (1743–94) published *Elementary Treatise on Chemistry* in 1789, and in it he included a table that listed thirty-one chemical elements.

Step Three — Atoms

Even if one were satisfied that the universe is made up of various elements, existing singly and in combination, and were convinced that the various elements had been correctly identified, isolated, and studied, the question of what the basic materials of the universe might be would remain not completely answered.

After all, if gold is an element, what is gold made of? Is it made of little particles of gold mashed together into bigger chunks? If so, how small a particle of gold can you have? Is there such a thing as an ultimately small gold particle that can be divided into nothing smaller?

Some Greek philosophers thought that every element consisted of tiny indivisible particles, and Democritus (470–380 B.C.) called such a particle *atomos* (*atom* in English) from a Greek word meaning *indivisible.*

Democritus' atomic theory did not carry conviction to the ancient Greeks, but once experimental science was established, the evidence in favor of atoms began slowly to accumulate.

The English chemist John Dalton (1766–1844) was the first to summarize all such evidence in a convincing manner. In 1808, he published a book entitled *New System of Chemical Philosophy* in which the atomic theory was spelled out in great detail.

The answer to the question, "What is the universe made of?" could then be given as "Atoms."

There were, of course, different kinds of atoms, one kind for each element. Gold was made up of gold atoms; iron of iron atoms; oxygen of oxygen atoms, and so on. Atoms could join in atom combinations called "molecules" and all the myriad of substances we see about us that are not elements are made up of molecules that, in turn, are made up of more than one kind of atom. Some of the substances in living organisms consist of molecules made up of millions of atoms of five or six different kinds.

Of course, if we stick to experimental science, it might well seem that we can only deal with those portions of the universe with which we could experiment; that is, the accessible portions of the earth itself. In that case, we might never really have an answer to the question, "What is the *universe* made of?" We could only hope that the matter about us is representative of the universe.

The French mathematician Auguste Comte (1798–1857) pointed this out in 1835 and suggested that the question of the chemical structure of the stars must remain forever unanswerable.

Two years after Comte died, however, the German physicist Gustav Robert Kirchhoff (1824–87) worked out the principles of spectroscopy, and by 1862, the Swedish physicist Anders Jonas Angstrom (1814–74) used it to show there was hydrogen in the sun.

Other spectroscopic investigations made it quite clear that the same atoms that existed about us were also to be found in the heavenly bodies. The atomic answer was indeed for the universe and not for earth alone.

Still, as the number of elements mounted and, therefore, the number of different kinds of atoms, scientists grew restive. Intuition seemed to indicate that the answer to the question "What is the universe made of?" had to be simple. If the basic building blocks were numerous and, apparently, unrelated, then surely that was the best sign that they weren't really basic but that they must in turn be composed of still more fundamental objects that *were* few in number.

By 1869 there were sixty-three different elements known, with no end in sight. In that year, however, the Russian chemist Dmitri Ivanovich Mendeleev (1834–1907) published his first version of the periodic table. In it, the elements were divided into families that were related in properties among themselves. This made the elements more orderly and not so miscellaneous and eased the discomfort of scientists somewhat.

Yet by the 1890s there were over eighty elements known and there was *still* no end in sight.

Step Four — Electrons and Atomic Nuclei

In the last quarter of the 1800s, scientists had been studying "cathode rays," produced when an electric current was forced through a vacuum. The studies made it seem that electricity, like matter, might be composed of indivisible units. The Irish physicist George Johnstone Stoney (1826–1911) suggested, in 1891, that the indivisible unit of electric charge be called an *electron.*

In 1897, the English physicist Joseph John Thomson (1856–1940) presented the final evidence that cathode rays consisted of electrically charged particles, and those received Stoney's name. Thomson was further able to show that the mass of the electron was only a small fraction ($1/1837$, actually) of the hydrogen atom, which was the lightest atom known. The electron was a "subatomic particle," the first to be discovered.

Could it be that electrons were purely a phenomenon of electric charge and had nothing to do with matter?

No! In 1896, the French physicist Antoine Henri Becquerel (1852–1908) had shown that uranium atoms broke down and gave off penetrating radiations. By 1900, some of those radiations were shown to consist of speeding electrons that must have emerged from the uranium atoms.

Nor was this a matter of one peculiar element. In 1902, the German physicist Philipp E. A. Lenard (1862–1947) showed that certain perfectly stable metals, when exposed to light, gave off electrons. Clearly, atoms were not indivisible but were made up of smaller particles still, including electrons.

The British physicist Ernest Rutherford (1871–1937) bombarded thin metal foil with radioactive radiation and, in 1911, produced evidence to show that atoms contained nearly all their mass in a very tiny atomic nucleus in the center of the atom. Whereas the typical atom was 10^{-8} centimeters in diameter, the atomic nucleus was 10^{-13} centimeters in diameter. The nucleus had only $1/100{,}000$ the diameter of the atom and $1/1{,}000{,}000{,}000{,}000{,}000$ the volume. The outer regions of the atom were filled with light electrons.

The English physicist Henry Gwyn-Jeffreys Moseley (1887–1915) was able to show, in 1913, that the atomic nuclei of different elements had positive electric charges of characteristic size —charges which were always the same for atoms of any one element, and always different for atoms of different elements.

It might seem, then, that the universe was made up of atomic nuclei differing among themselves in the size of their positive electric charge, plus electrons, all of the same negative charge, surrounding each nucleus in just large enough numbers to match the nuclear charge and produce a neutral atom.

By 1916, the American chemist Gilbert Newton Lewis (1875–1946) began the process of showing the electrons to exist in concentric shells within the atom and using that to explain chemical properties. This accounted for the existence of families of elements and explained why the periodic table existed in the form it did.

What's more, the existence of charges of fixed values on the atomic nuclei limited the number of elements that could exist. It became clear that there couldn't be any more than about eighty different stable elements.

Step Five — Electrons, Protons, and Neutrons

The atomic nuclei were not entirely satisfactory as basic constituents of the universe. Where electrons were all alike, atomic nuclei differed among themselves both in mass and in electric charge. The smallest nucleus, that of hydrogen, had a positive electric charge equal in size to the negative charge on the electron, but all other nuclei had positive charges that were integral multiples of that on the hydrogen nucleus. It seemed reasonable to suppose, then, that the atomic nucleus consisted of varying numbers of whatever particle it was that made up the hydrogen nucleus. Rutherford called the hydrogen-nucleus particle the *proton* from the Greek word for *first*.

To begin with, though, it was quite clear that the atomic nucleus could not exist of protons only. For instance, the helium nucleus had twice the charge of the hydrogen nucleus, but four times the mass. It took four protons to supply the mass, but they would supply four times the charge, too.

For some years, it was thought that electrons were also present in nuclei and served to neutralize some of the positive charge. Unfortunately, the protons and electrons also had something called spin and so did the nuclei. If protons and electrons were so arranged as to account for the mass and charge of the nucleus, they often did not account for the spin.

Then, in 1932, the English physicist James Chadwick (1891–1974) discovered the neutron, which was just about as massive as the proton but did not carry an electric charge. It became clear that atomic nuclei were made up of protons and neutrons. Combinations of protons and neutrons could be made to account for the mass and charge of all the nuclei, and the spins as well.

Such combinations also explained the existence of isotopes, first demonstrated in 1914 by the English chemist Frederick Soddy (1877–1956). Though all nuclei of a given element had the same number of protons, they could be divided into two or more groups, each with a slightly different number of neutrons.

For a few happy years, it seemed that the universe was made up of just three different kinds of particles: electrons with a charge of −1, protons with a charge of +1, and neutrons with a charge of 0. What could be neater?

Step Six — Leptons and Hadrons

Euphoria did not last long.

For one thing, there was a curious asymmetry. Protons and electrons had electric charges of precisely the same size though of opposite nature, but the proton was 1,836 times as massive as the electron. Why?

An answer of sorts came in 1930, when the English physicist Paul

A. M. Dirac (1902–) pointed out that each particle should have an "antiparticle," equal and opposite. In 1932, the American physicist Carl David Anderson (1905–), studying cosmic ray particles, detected the *antielectron,* or positron, in the debris. It had a mass and charge equal to the electron, but the charge was positive rather than negative. In 1955, the Italian-American physicist Emilio Segrè (1905–) and the American physicist Owen Chamberlain (1920–) detected the *antiproton,* which had a mass and charge equal to the proton, but with a negative rather than a positive charge.

It was clear that there were two parts to the universe, so to speak, an ordinary part and an antipart, and that each was asymmetric but in mirror-image form. The two together formed a symmetry, but at the price of doubling the complexity of the universe.

Other complications were developing, too. There was, for instance, a puzzle over the nature of the forces holding the atomic nucleus together.

As late as 1932, it had seemed that two forces in the universe were sufficient to explain the motions and interactions of all its parts: the gravitational interaction and the electromagnetic interaction.

The electromagnetic interaction was far, far the stronger of the two, but the gravitational interaction dominated the universe as a whole because it was entirely an attractive force, while the electromagnetic interaction involved both attractions and repulsions that largely neutralized each other.

Once it was discovered that the atomic nucleus consisted of protons and neutrons, neither interaction would explain its existence. The gravitational interaction was far, far too weak to hold it together, while the electromagnetic interaction acted to drive it apart. There had to be some nuclear interaction stronger than the electromagnetic interaction to hold it together against the force of electromagnetic repulsion. And it had to be short-range so as not to be noticeable at more than nuclear distances.

In 1935, the Japanese physicist Hideki Yukawa (1907–82) worked out the theoretical background for such a "strong interaction." Some time later, the Italian physicist Enrico Fermi (1901–54) showed that a second nuclear interaction, a much weaker one (the "weak interaction"), was needed to account for radioactive breakdown and a number of other particle interactions.

It then became possible to divide subatomic particles into two types. There are the *hadrons* (from a Greek word meaning *thick* or *strong*), which can respond to the strong interaction as well as to the weak; and the *leptons* (from a Greek word meaning *weak*), which can respond only to the weak interaction and *not* to the strong.

The proton, antiproton, neutron, and antineutron are hadrons, while the electron and antielectron are leptons.

Another source of complication came about when scientists studied nuclear interactions in increasing detail and found that additional particles were needed if all the events were to be explained. In 1931, the Austrian physicist Wolfgang Pauli (1900–58) had suggested that when an electron was emitted in a radioactive breakdown, it had to be accompanied by another particle without either mass or electric charge. Fermi called it a *neutrino* (*little neutral one* in Italian).

A particle without mass or charge is difficult indeed to detect and it wasn't until 1956 that the task was accomplished by the American physicist Frederick Reines (1918–). Naturally, it turned out that there was not only a neutrino but an antineutrino.

Then, too, as scientists dealt with higher and higher energies, either when studying cosmic rays or by working with more and more elaborate particle accelerators, they found that more and more energetic particles could be formed and detected. These were all unstable particles quickly breaking down to more stable ones such as electrons or protons, but they existed, even if only temporarily, and they complicated the universe.

Thus, in 1935, Carl Anderson detected the *muon*, with properties identical to those of the electron but possessing 207 times its mass. There is also an antimuon. In addition there are muon neutrinos and muon antineutrinos which are identical to the ordinary electron-neutrinos and antineutrinos in every respect that we can measure but which behave differently in nuclear reactions and so must have *some* difference we're not subtle enough yet to see.

The muons and their neutrinos are also leptons and there are indications that there are still more massive leptons, a *tau electron* with its associated antiparticle together with tau neutrinos and tau antineutrinos. If the energy supply were unlimited, it might be that there would be an endless series of leptons of ever greater mass, each with its antiparticle, neutrino and antineutrino.

The hadrons exist in even greater variety, some less massive than the proton and neutron, some more massive. The less massive ones are *mesons*, from the Greek word for *intermediate*, because their mass is intermediate between the proton and electron. The more massive ones are *hyperons* from the Greek word for *beyond.*

The least massive hadron is the *pion,* first detected in 1947 by the English physicist Cecil Frank Powell (1903–69). It is about one-seventh as massive as a proton (270 times as massive as an electron) and comes in five varieties. There is a positive and a negative pion, each with its antiparticle, and a neutral pion which is its own antiparticle.

Mesons and hyperons multiplied rapidly until well over a hundred had been discovered, with more constantly piling in. The sheer number of hadrons called for some explanation. Could they all be built up of still more basic particles?

Step Seven—Leptons and Quarks

In 1953, the American physicist Murray Gell-Mann (1929–) worked out a system whereby the hadrons could be arranged in families. It was a sort of periodic table of hadrons like Mendeleev's table of chemical elements. To make sense of Mendeleev's table one needed three subatomic particles, the electron, proton, and neutron. To make sense of Gell-Mann's periodic table of hadrons, one needed three subhadronic particles.

Gell-Mann called his subhadronic particles *quarks,* from a phrase in James Joyce's *Finnegans Wake,* which goes "three quarks for Muster Mark."

Gell-Mann needed only two types of quarks to begin with, and they are called *up-quarks* and *down-quarks,* or *u* and *d,* for purposes of distinction, though the description mustn't be taken literally. The electric charge of the *u* is +⅔ and that of the *d* is −⅓. Two *d*'s and a *u* total 0 and make a neutron. Two *u*'s and a *d* total +1 and make a proton. Naturally there are anti-*d*'s and anti-*u*'s and these can form the antineutron and antiproton.

Other combinations form various hyperons. If the quarks are taken two at a time, a quark and an antiquark, mesons are formed. Whatever the combinations, the fractional charges must disappear. The overall charge of quark combinations must be 0, 1, 2 . . .

There seem to be analogies between leptons and quarks. Just as in leptons there is a basic pair, electron/neutrino and the antiparticles of each; so in quarks there is a basic pair *u*-quark/*d*-quark and the antiparticles of each.

In leptons, additional energy can produce electron analogs of more and more mass—muons, tau electrons and so on, each with its neutrinos and its antiparticles. In quarks, additional energy can produce quarks of more and more mass, each with its pair and its antiparticles.

Thus, more energetic than the *u* and the *d* are the *s*-quark and the *c*-quark (where the *s* and *c* are stated, with physicists' whimsy to stand for *strangeness* and *charm*). Beyond that may be the *t*-quark and the *b*-quark (*top* and *bottom,* or, to be more poetic, *truth* and *beauty*), and so on. Each level is a *flavor.*

The world of quarks is considerably more complicated than the world of leptons, however. The leptons are distinguished among themselves by mass and charge, and so are the quarks—but the quarks are also distinguished among themselves by properties that leptons do not possess but which are called (metaphorically only) *color.* Each different flavor of quark comes in varieties which are called *red, blue,* and *green.*

When quarks get together three at a time, there must be one red quark, one green quark, and one blue quark, the combination being without color, or *white.* When they get together two at a time, it is always a color and an anticolor. The colors always disappear in the quark combinations, as the fractional charges do. The study of quark combinations is therefore called *quantum chromodynamics,* or *QCD,* the *chromo-* coming from the Greek for *color.*

What's more, the quarks *do* combine, while leptons do not, since the quarks are subject to the strong interaction and the leptons are not. Involved in the quark combination is a special particle which is constantly exchanged by them and which holds them together. This is the *gluon,* so called for obvious reasons.

So far, no one has been able to pull hadrons apart and study the individual quarks, and there are some theories of quark combinations that hold it is impossible to do that.

Another alternative is to form quarks from scratch by concentrating enough energy into a small volume, as by smashing together very energetic streams of electrons and antielectrons. The quarks produced would instantly combine into hadrons and antihadrons which would stream off in opposite directions. If there were *enough* energy there would be three streams forming a three-leaf clover: hadrons, antihadrons, and gluons. The two-leaf clover has been formed and in 1979, there were announcements of experiments in which a very rudimentary third leaf was just beginning to form. This is considered a confirmation of the quark theory.

Step Eight—?

The lepton-quark theory is the best way we have yet of explaining the fundamental basis of the universe, but there are still questions. Why are there both quarks *and* leptons? Why are there quarks in so many colors? Why must quarks combine while leptons can remain free?

Could there be something more basic still?

An Israeli physicist, Haim Harari suggests subleptonic and subquarkic particles he calls *rishons* from a Hebrew word meaning *first.* He suggests a T-rishon with an electric charge of +⅓ and a V-rishon with an electric charge of 0 and antirishons in both cases, and contends that leptons and quarks can *both* be built up out of rishons taken three at a time.

Is he correct? And if he is correct, will this be the final end? Will we have gotten down to basics at last? Or can it be that there are no basics, but that we are sliding down a chute without a bottom and that the search for fundamental particles leads us on toward a goal that recedes as fast as we approach it?

11★ AND AFTER MANY A SUMMER DIES THE PROTON

If any of you aspire to the status of Very Important Person, let me warn you sulkily that there are disadvantages. For myself, I do my best to avoid VIP-dom by hanging around my typewriter in a state of splendid isolation for as long as possible. And yet—the world intrudes.

Every once in a while, I find myself slated to attend a grand function at some elaborate hotel, and the instructions are "black tie." That means I've got to climb into my tuxedo. It's not really very difficult to do so and once I'm inside it, with the studs and links in place, with the tie hooked on and the cummerbund adjusted, I don't feel very different. It's just the principle of the thing. I'm not a tuxedo person; I'm a baggy-old-clothes person.

Just the other night I was slated to appear, tuxedo-ablaze-in-glory at the Waldorf-Astoria. I had been invited—but I had not received any tickets.

Whereupon I said to Janet (who made her usual wifely suggestion that she seize her garden shears and cut great swatches out of my luxuriant sideburns and received my usual husbandly refusal), "Listen, if we get there and they won't let us in without tickets, please don't feel embarrassed. We'll just leave our coats in the checkroom, go down two flights to the Peacock Alley and eat there."

In fact, I was hoping we'd be turned away. Of all the restaurants I've tried in New York, the Peacock Alley is my favorite. The closer we got to the hotel, the more pleasant was my mind's-eye picture of myself wreaking havoc with the comestibles at the Peacockian festive board.

Finally, there we were, standing before a group of fine people who barred the way to the Grand Ballroom, with instructions to keep out the riffraff.

"I'm sorry," I said, firmly, "but I don't have any tickets."

Whereupon a clear whisper sounded from one young woman on the other side of the table, "Oh, my goodness! Isaac Asimov!"

And instantly, Janet and I were hustled into the VIP room and my hopes for the Peacock Alley went a-glimmering.*

So let us turn, by an easy progression of thought, to that VIP of the subatomic particles: the proton.

Fully 90 percent of the mass of that portion of the universe of which we are most aware—the stars—consists of protons. It is therefore apparently fair to say that the proton is the very stuff of the universe and that if anything deserves the rating Very Important, it is the proton.

Yet the proton's proud position on the throne of subatomic VIPdom is now being shaken.

In the first place, there is the possibility (see chapter 16) that it is not the proton after all that is the stuff of the universe, but the neutrino, and that the proton makes up only a very inconsiderable portion of the universal mass.

In the second place, it is possible that the proton is not even im-

* It was all right. It was a very good banquet and a lot of fun.

mortal, as has long been thought, but that after many a summer each one of the little things faces decay and death even as you and I.

But let's start from the beginning.

At the moment, there seem to be two fundamental varieties of particle: leptons and quarks (see the previous chapter).

There are different sorts of leptons. First, there are the electron, the muon, and the tauon (or tau electron). Then there are the mirror-image particles, the antielectron (or positron), the antimuon, and the antitauon. Then there is a neutrino associated with each of the above: the electron neutrino, the muon neutrino, and the tauon neutrino, plus, of course, an antineutrino for each.

That means twelve leptons altogether that we know of, but we can simplify the problem somewhat by ignoring the antiparticles, since what we have to say about the particles will hold just as firmly for the antiparticles. Furthermore, we will not try to distinguish between the neutrinos since there is a chance that they may oscillate and swap identities endlessly (as I shall explain later in the book).

Therefore let us speak of four leptons—the electron, the muon, the tauon, and the neutrino.

Different particles have different rest masses. For instance, if we set the rest mass of the electron at 1, the rest mass of the muon is about 207, and that of the tauon is about 3,600. The rest mass of the neutrino, on the other hand, may be something like 0.0001.

Mass represents a very concentrated form of energy and the general tendency seems to be for massive particles to change, spontaneously, into less massive particles.

Thus, tauons tend to break down into muons, electrons, and neutrinos and to do it quickly, too. The half-life of a tauon (the period of time during which half of them will have broken down) is only about five trillionths of a second (5×10^{-12} seconds).

Muons, in turn, break down to electrons and neutrinos, but since muons are less massive than tauons they seem to last a bit longer and have half-lives of all of 2.2 millionths of a second (2.2×10^{-6} seconds).

You might expect that electrons, then, might live a little longer still, and break down to neutrinos, and that neutrinos, after a perhaps quite respectable lifetime, might melt away to complete masslessness, but that's not the way it works.

Leptons can't disappear altogether, provided we are dealing with particles only or antiparticles only, and not a mixture of the two. An electron and an antielectron can combine and mutually annihilate, converting themselves into zero-mass photons (which are not leptons), but that's another thing and we're not dealing with it.

As long as we have only particles (or only antiparticles), leptons must remain in existence; they can shift from one form to another, but cannot disappear altogether. That is the *law of conservation of lepton number,* which also means that a lepton cannot come into existence out of a nonlepton. (A lepton *and* its corresponding antilepton can simultaneously come into existence out of nonleptons, but that's another thing.) And don't ask *why* lepton number is conserved; it's just the way the universe seems to be.

The conservation of lepton number means that the neutrino, at least, should be immortal and should never decay, since no still-less-massive lepton exists for it to change into. This fits the facts, as nearly as we can tell.

But why should the electron be stable, as it seems to be? Why doesn't it break down to neutrinos? That would not violate the law of conservation of lepton number.

Ah, but leptons may possess another easily measurable characteristic—that of electric charge.

Some of the leptons, the various neutrinos and antineutrinos, have no electric charge at all. The others—the electron, the muon, and the tauon—all have an electric charge of the same size, which, for historical reasons, is considered to be negative and is usually set equal to unity. Each electron, muon and tauon has an electric charge of -1; while every antielectron, antimuon, and antitauon has an electric charge of $+1$.

As it happens, there is a *law of conservation of electric charge,* which is a way of saying that electric charge is never observed to disappear into nothing, or appear out of nothing. No lepton decay can affect the electric charge. (Of course, an electron and an antielectron can interact to produce photons, and the opposite charges, $+1$ and -1, will cancel. What's more, a lepton and an antilepton can be formed simultaneously, producing both a $+1$ and a -1 charge where no charge existed before—but these are different things from those we are discussing. We are talking about particles and antiparticles as they exist separately.)

The least massive of the leptons with charge is the electron. That means that though more massive leptons can easily decay to the electron, the electron cannot decay because there is nothing less massive which can hold an electric charge, and that electric charge *must* continue to exist.

To summarize, then:

Muons and tauons can come into existence under conditions where the general energy concentration is locally very high, say, in connection with particle accelerators or cosmic ray bombardment; but once formed, they cannot last for long. Under ordinary conditions, removed from high-energy events, we would find neither muons nor tauons and the universal content of leptons is restricted to the electron and the neutrino. (Even the antielectron does not exist in significant numbers for reasons to be taken up in chapter 14.

Let us pass on next to the other basic variety of particle, the quark. Quarks, like leptons, exist in a number of varieties, but with a number of important differences.

For one thing, quarks carry fractional electric charges, such as $+2/3$ and $+1/3$. (Antiquarks have charges of $-2/3$ and $-1/3$, naturally.)

Furthermore, the quarks are subject to the "strong interaction," which is enormously more intense than the "weak interaction" to which leptons are subject. The intensity of the strong interaction makes it unlikely (even, perhaps, impossible) for quarks to exist in isolation. They seem to exist only in bound groups that form according to rules briefly described in the previous chapter. One very common way of grouping is to have three quarks associate in such a way that the overall electric charge is either 0, 1, or 2 (positive in the case of some, negative in the case of others).

These three-quark groups are called *baryons,* and there are large numbers of them.

Again, however, the more massive baryons decay quickly into less massive baryons, which decay into still less massive baryons, and so on. As side products of this decay, mesons are produced which are particles made up of only two quarks. There are no stable mesons. All break down more or less rapidly into leptons; that is, into electrons and neutrinos.

There is, however, a *law of conservation of baryon number,* so that whenever a baryon decays, it must produce another baryon, whatever else it produces. Naturally, when you get to the baryon of the lowest possible mass, no further decay can take place.

The two baryons of lowest mass are the proton and the neutron, so that any other baryon, of the many dozens that can exist, quickly slides down the mass scale to become either a proton or a neutron.

These two baryons are the only ones that exist in the universe under the ordinary conditions that surround us. They tend to combine in varying numbers to form the atomic nuclei.

The proton and neutron differ, most obviously in the fact that the proton has an electric charge of +1, while that of the neutron is 0. Naturally, atomic nuclei, which are made up of protons and neutrons, all carry a positive electric charge of a quantity equal to the number of protons present. (There are also such things as antiprotons with a charge of —1, and antineutrons which differ from neutrons in magnetic properties, and these can group together to form negatively charged nuclei and antimatter, but never mind that right now.)

The positively charged nuclei attract negatively charged electrons in numbers that suffice to neutralize the particular nuclear charge, thus forming the different atoms with which we are familiar. Different atoms, by transferring or sharing one or more electrons, form molecules.

But the proton and neutron differ slightly in mass, too. If we call the electron's mass 1, then the proton's mass is 1,836 and the neutron's mass is 1,838.

When the two exist in combination in nuclei, they tend to even out their properties and to become, in effect, equivalent particles. Inside nuclei, then, they can be lumped together and referred to as *nucleons*. The entire nucleus is then stable, although there are nuclei where the proton-neutron mixture is not of the proper ratio to allow a perfect evening out of properties, and which are therefore radioactive—but that's another story.

When the neutron is in isolation, however, it is not stable. It tends to decay into the slightly less massive proton. It emits an electron, which carries off a negative charge, leaving a positive charge behind on what had been a neutron. (This simultaneous production of a negative *and* a positive charge does not violate the law of conservation of electric charge.) A neutrino is also formed.

The mass difference between proton and neutron is so small that the neutron doesn't decay rapidly. The half-life of the isolated neutron is about twelve minutes.

This means that the neutron can exist for a considerable length of time only when it is in combination with protons, forming an atomic nucleus. The proton, on the other hand, can exist all by itself for indefinite periods and can, all by itself, form an atomic nucleus, with a single electron circling it—forming the ordinary hydrogen atom.

The proton is thus the only truly stable baryon in existence. It, along with the electron and the neutrino (plus a few neutrons that exist in atomic nuclei), makes up virtually all the rest mass of the universe. And since protons outshine the others in either number or individual rest mass, the proton makes up 90 percent of the mass of such objects as stars. (The neutrinos may be more massive, in total, but they exist chiefly in interstellar space.)

Consider the situation, however, if matters were the other way around and if the neutron were slightly less massive than the proton. In that case, the proton would be unstable and would decay to a neutron, giving up its charge in the form of a positively charged antielectron (plus a neutrino). The antielectrons so formed would annihilate the electrons of the universe, together with the electric charge of both, and left behind would be only the neutrons and neutrinos. The neutrons would gather, under the pull of their overall gravitational field, into tiny neutron stars, and those would be the sole significant structures of the universe.

Life as we know it, would, of course, be utterly impossible in a neutron-dominated universe, and it is only our good fortune that the proton is slightly less massive than the neutron, rather than vice versa, that gives us expanded stars, and atoms—and life.

Everything, then, depends on the proton's stability. How stable is it? Our measurements show no signs of proton decay, but our measurements are not infinitely delicate and precise. The decay might be there but might be taking place too slowly for our instruments to catch it.

Physicists are now evolving something called the Grand Unified Theory (GUT) by which one overall description will cover the electromagnetic interaction (affecting charged particles), the weak interaction (affecting leptons), and the strong interaction (affecting quarks and quark groupings such as mesons, baryons, and atomic nuclei).

According to GUT, each of the three interactions is mediated by *exchange particles* with properties dictated by the necessity of making the theory fit what is already known. The electromagnetic exchange particle is the photon, which is a known particle and very well understood. In fact, the electromagnetic interaction is well described by quantum electrodynamics, which serves as a model for the rest of the GUT.

The weak interaction is mediated by three particles symbolized as W^+, W^-, and Z^0, which have not yet been detected. The strong interaction is mediated by no less than eight "gluons," for whose existence there is reasonable evidence, albeit indirect.

The more massive an exchange particle is, the shorter its range. The photon has a rest mass of zero, so electromagnetism is a very long-range interaction and falls off only as the square of the distance. (The same is true of the gravitational interaction, which has the zero-mass graviton as the exchange particle, but the gravitational interaction has so far resisted all efforts to unify it with the other three.)

The weak-exchange particles and the gluons have considerable mass, however, and therefore the intensity of their influence falls off so rapidly with distance that that influence is measurable only at distances comparable in size to the diameter of the atomic nucleus, which is only a tenth of a trillionth of a centimeter (10^{-13} centimeters) across or so.

GUT, however, in order to work, seems to make necessary the existence of no fewer than twelve more exchange particles, much more massive than any of the other exchange particles, therefore extremely short-lived and difficult to observe. If they *could* be observed, their existence would be powerful evidence in favor of GUT.

It seems quite unlikely that these ultramassive exchange particles can be directly detected in the foreseeable future, but it would be sufficient to detect their effects, if those effects were completely unlike those produced by any other exchange particles. And such an effect does (or, at any rate, *might*) exist.

If one of these hypermassive exchange particles should happen to be transferred from one quark to another within a proton, a quark would be changed to a lepton, thus breaking both the law of conservation of baryon number and the law of conservation of lepton number. The proton, losing one of its quarks, becomes a positively charged meson that quickly decays into antielectrons, neutrinos, and photons.

The hypermassive exchange particles are so massive, however, that their range of action is roughly 10^{-29} centimeters. This is only a tenth of a quadrillionth (10^{-16}) the diameter of the atomic nucleus. This means that the point-sized quarks can rattle around inside a proton for a long, long time without ever getting sufficiently close to one another to exchange a proton-destroying exchange particle.

In order to get a picture of the difficulty of the task of proton decay, imagine that the proton is a hollow structure the size of the planet earth, and that inside that vast planetary hollow are exactly three objects, each about a hundred-millionth of a centimeter in diameter—in other words, just about the size of an atom in our world. Those "atoms" would have diameters that represent the range of action of the hypermassive exchange particles.

These "atoms," within that earth-sized volume, moving about randomly, would have to collide before the proton would be sent into decay. You can easily see that such a collision is not likely to happen for a long, long time.

The necessary calculation makes it seem that the half-life for such proton decay is ten million trillion trillion years (10^{31} years). After many a summer, in other words, dies the proton—but after many, many, *many* a summer.

To get an idea of how long a period of time the proton's half-life is, consider that the lifetime of the universe to this point is usually taken as 15,000,000,000 years—fifteen billion in words, 1.5×10^{10} years in exponential notation.

The expected lifetime of the proton is roughly 600 million trillion (6×10^{20}) times that.

If we set the mighty life of the universe as the equivalent of one second, then the expected half-life of the proton would be the equivalent of 200 trillion years. In other words, to a proton, the entire lifetime of the universe so far is far, far less than an eye blink.

Considering the long-lived nature of a proton, it is no wonder that its decay has not been noted and that scientists have not detected the breakage of the laws of conservation of baryon number and lepton number and have gone on thinking of those two laws as absolutes.

Might it not be reasonable, in fact, to ignore proton decay? Surely a half-life of 10^{31} years is so near to infinite, in a practical sense, that it might as well be taken as infinite and forgotten.

However, physicists can't do that. They must try to measure the half-life of proton decay, if they can. If it turns out to be indeed 10^{31} years, then that is powerful support for GUT; and if it turns out that the proton is truly stable then GUT is invalid or, at the very least, would require important modification.

A half-life of 10^{31} years doesn't mean that protons will all last for that long and then just as the last of those years elapses, half of them will decay at once. Those atom-sized objects moving about in an earth-sized hollow could, by the happenstance of random movement, manage to collide after a single year of movement, or even a single second. They might, on the other hand, just happen to move about for 10^{100} or even $10^{1,000}$ years without colliding.

Statistically, though, since there are many, many protons, some decays should take place all the time. In fact, if the half-life of the proton were merely ten thousand trillion years (10^{16} years) there would be enough proton decays going on within our bodies to kill us with radioactivity.

Even with a half-life of 10^{31} years, there would be enough proton decays going on right now to destroy something like thirty thousand trillion trillion trillion protons (3×10^{40}) *every second* in the universe as a whole, or three hundred thousand trillion trillion (3×10^{29}) every second in our galaxy alone, or three million trillion (3×10^{18}) every second in our sun alone, or three thousand trillion (3×10^{15}) every second in Jupiter alone, or three billion (3×10^{9}) every second in earth's oceans.

This begins to look uncomfortably high, perhaps. Three billion proton decays every second in our oceans? How is that possible with an expected lifetime so long that the entire life of the universe is very nearly nothing in comparison?

We must realize how small a proton is and how large the universe is. Even at the figures I've given above it turns out that only enough protons decay in the course of a billion years throughout the entire universe to be equivalent to the mass of a star like our sun. This means that in the total lifetime of our universe so far, the universe has lost through proton decay the equivalent of fifteen stars the mass of the sun.

Since there are 10,000,000,000,000,000,000,000 (ten billion trillion, or 10^{22}) stars in the universe as a whole, the loss of fifteen through proton decay can easily be ignored.

Put it another way. In one second of the hydrogen fusion required to keep it radiating at its present rate, the sun loses six times as much mass as it has lost through proton decay during the entire five-billion-year period during which it has been shining.

The fact that, despite the immensely long half-life of the proton, decays go on steadily at all times, raises the possibility of the detection of those decays.

Three billion decays every second in our oceans sounds as though it should be detectable—but we can't study the ocean as a whole with our instruments and we can't isolate the ocean from other possibly obscuring phenomena.

Nevertheless, tests on considerably smaller samples have fixed the half-life of the proton as no shorter than 10^{29} years. In other words, experiments have been conducted where, if the proton's half-life were shorter than 10^{29} years, protons would have been caught in the act of decaying—and they weren't. And 10^{29} years is a period of time only one one-hundredth the length of 10^{31} years.

That means that our most delicate detecting devices combined with our most careful procedures need only be made a hundred times more delicate and careful in order just barely to detect the actual decay of protons if the GUT is on the nose. Considering the steady manner in which the field of subatomic physics has been advancing this century, this is a rather hopeful situation.

The attempt is being made, actually. In Ohio, the necessary apparatus is being prepared. Something like ten thousand tons of water will be gathered in a salt mine deep enough in the earth to shield it from cosmic rays (which could produce effects that might be confused with those arising from proton decay).

There would be expected to be 100 decays per year under these conditions, and a long meticulous watch *may,* just possibly *may,* produce results that will confirm the Grand Unified Theory and take us a long step forward indeed in our understanding of the universe.

12 ★ LET EINSTEIN BE!

Every once in a while I review books. I hate doing so because I hate being a critic; I hate reading with a view to making judgments or poking holes. I don't even know if I'm equipped to make judgments and poke holes. I just want to read for pleasure and profit, and continue the reading or stop it according to whether that p and p is there or not.

Every once in a while, though, someone asks me to do a review under conditions where I can't refuse, and in this case, I found myself with three books dealing with relativity. I did the job—but I also did some thinking.

Writing books that explain relativity to the layman is virtually big business. The theory of relativity is over seventy-five years old and it still needs explaining.

It is accepted by scientists; in fact, you can't understand modern physics without it. Yet the resistance to its concepts (never mind its mathematics) on the part of the layman never ebbs. Why are the attempts to explain relativity, while apparently endless, also apparently useless?

In this connection, consider a very famous epitaph intended for Isaac Newton, written by Alexander Pope (1688–1744):

Nature and Nature's laws lay hid in night:
God said, "Let Newton be!" and all was light.

Very true! And just as true, at least in popular estimation, are the two lines added in recent decades by the British journalist John C. Squire (1884–1958). These are:

It did not last: the Devil howling, "Ho!
Let Einstein be!" restored the status quo.

And there you are! Einstein advanced crazy ideas that violated common sense, that could not be absorbed or grasped, and the public will have none of it!

And yet, in certain ways, humanity has lived through such intellectual crises before. Einstein's relativity is not the first interpretation of the universe to violate common sense. It's just that earlier ones have ground their way into popular acceptance while relativity hasn't and perhaps never will.

I would like to give an instance of a situation as odd as anything Einstein ever dreamed up and yet one that *was* accepted.

Let us suppose that two people, Smith and Jones, are standing at point X along with you and me. Smith sets off in any direction at random, walking in a perfectly straight line at a steady 5 kilometers an hour. Jones sets off in the precisely opposite direction, also walking in a perfectly straight line at the same speed.

Let us suppose that neither Smith nor Jones requires food or drink; that neither gets tired; that neither encounters any obstacle such as mountains, deserts, or oceans; that neither deviates in any way from the straight line travel at 5 kilometers an hour.*

Let us suppose, further, that you and I remain at point X and have the ability, at every moment, to determine the distance of Smith and Jones from ourselves and from each other. Furthermore,

* This is a "thought experiment" and we are allowed to simplify matters by omitting all nonessential entanglements.

though we know nothing about geography, we are well versed in ordinary arithmetic and Euclidean geometry.

At the end of 1 hour, Smith is 5 kilometers away in one direction, Jones is 5 kilometers away in the opposite direction, and they are 10 kilometers apart.

At the end of 2 hours, Smith is 10 kilometers away in one direction, Jones is 10 kilometers away in the opposite direction, and they are 20 kilometers apart.

Since we know arithmetic we feel safe in predicting that at the end of 10 hours, Smith will be 50 kilometers away in one direction, Jones will be 50 kilometers away in the other direction and they will be 100 kilometers apart. (And sure enough, if we check the situation at the end of 10 hours, we will find our prediction was correct.)

Knowing Euclidean geometry, we know that a straight line can be extended indefinitely, and since Smith and Jones are walking in opposite directions on a straight line, we can extrapolate our arithmetic forever.

For instance, after 8,000 hours (that's very nearly a year), Smith will be 40,000 kilometers away in one direction, and Jones will be 40,000 kilometers away in the other direction and they will be 80,000 kilometers apart.

This can be continued indefinitely by ordinary arithmetic, ordinary geometry, and ordinary common sense. Anyone who would argue with such figures would have to be out of his or her mind.

Except that now we will set up a ridiculous assumption.

Let us suppose that 20,000 kilometers is an absolutely maximum separation.* No matter how long Smith walks away from us in a straight line in a given direction, he will never get more than 20,000 kilometers away from us. What's more, no matter how long Jones walks away from us in a straight line in the opposite direction, *he* can never get more than 20,000 kilometers away from us either. What's still more, Smith and Jones, as they firmly march off in opposite directions, can never get more than 20,000 kilometers away from each other.

Some hardheaded no-nonsense guy would surely object.

"This is simply insane!" Hardhead would say. "Why twenty thousand kilometers?"

We shrug. That's just the way things are. It's our assumption.

* A better figure would be 20,037 kilometers, but I trust you will permit me the convenient approximation.

"All right, then," says Hardhead, "suppose Smith has walked onward until he's twenty thousand kilometers from us, and then suppose he keeps on walking. Doesn't he *have* to get farther away from us?"

No! If he reaches that maximum distance and insists on keeping on walking in the same straight line, he *can't* get farther from us. The distance between us changes but only in the sense that he now gets *closer* to us. After all if he can't get farther when he moves, he must get closer—and he just can't get farther.

"Well, that's the most exasperatingly stupid rule anyone has ever thought up," says Hardhead, "and you must be a prize idiot to dream it up."

Well, perhaps, but let's see what the consequences are. Smith is walking in one direction, Jones in the other. After 2,000 hours, Smith is 10,000 kilometers in one direction, and Jones is 10,000 kilometers in the other direction, and they are 20,000 kilometers apart. Right?

"Right," says Hardhead.

If they keep on walking, Smith continues to get farther away from us and Jones continues to get farther away from us in the other direction, *but they can't get farther away from each other,* because they have reached the 20,000-kilometer maximum separation. As they keep on walking, by the rules we have set up, though each gets steadily farther away from us, each gets as steadily closer to the other.

After 3,000 hours, Smith has added 5,000 kilometers to his distance from us, and so has Jones. Each is now 15,000 kilometers away from us in opposite directions. Each, however, has *decreased* the distance from the other by 5,000 kilometers, so that they are now 10,000 kilometers apart.

"Let's get this straight," says Hardhead. "Smith is fifteen thousand kilometers away from us in one direction, and Jones is fifteen thousand kilometers away from us in the other direction, but Smith is only ten thousand kilometers from Jones. You're telling me that fifteen thousand plus fifteen thousand equals ten thousand. Do you realize that if you carry on with this insanity, Smith and Jones will *meet?"*

Exactly! After 4,000 hours, Smith has traveled 20,000 kilometers in one direction, and Jones has traveled 20,000 kilometers in the other direction, and they will meet at point Y.

"So that twenty thousand plus twenty thousand equals zero," says Hardhead. "That's rich, that is. And what if they keep on walking?"

Well, they started off facing and walking in opposite directions and when they meet they are still facing and walking in opposite directions. If they keep on, they will pass each other. Smith retraces Jones's steps; Jones retraces Smith's steps. They begin to move farther from each other but, having passed the 20,000-kilometer mark, they each begin to move closer to us.

After 6,000 hours, Smith and Jones have gotten halfway back to us and are each only 10,000 kilometers away from us, but they have been moving away from each other and are now 20,000 kilometers apart. After they have reached that maximum separation, they start moving toward each other again. After 8,000 hours, Smith and Jones face each other once again and both are zero kilometers away from us. We are all together again.

"I see. I see," says Hardhead. "Smith travels steadily in a given direction for eight thousand hours, and Jones travels steadily in the opposite direction for eight thousand hours. Neither one veers from the original direction and yet they end up after all that walk right back home again."

Yes, indeed, and if they keep on walking they will meet again at Y, and then at X, and then at Y, and so on forever. And through all eternity they will never be more than 20,000 kilometers from their starting point.

What's more they can do this in any direction. Smith can move away from the original point X in a totally different direction from the one he first chose, and Jones can move in the direction opposite to that and they will still meet at point Y. It will be the same point Y, no matter which straight line they move in opposite directions upon.

"The *same* point Y? How can you tell that?"

It follows from the original assumptions. Suppose that by going along two separate lines they meet at point Y in the first case and point Y′ in the second. Both points would have to be 20,000 kilometers from us by the original assumption.

Suppose, then, we try to walk from point Y to point Y′. Since point Y is 20,000 kilometers away from us and we can't walk farther away by our original assumption, then no matter which direction we take we get closer to us than 20,000 kilometers. When we arrive at point Y′ we are less than 20,000 kilometers from us, and yet point Y′ is 20,000 kilometers from us. The only way we can avoid the paradox is to suppose that point Y and point Y′ are identical.

In fact, if you study the situation further it turns out that, given our assumption of maximum separation, any two straight lines on the surface of the earth intersect at two different points 20,000 kilometers apart, even though Euclidean geometry tells us that two straight lines can intersect at only one point no matter how far they are extended.

Since all straight lines intersect, there are no parallel lines under the maximum-separation assumption.

Furthermore, without going into the details of the demonstrations which would be very convoluted, it could be shown that the shortest distance between two points where one is due west of the other is *not* along a line that goes due west.

In a place like the United States, to go from one point to another, which is due west, along the shortest line, one must head off a little north of west. (If you live in Australia you have to head off a little south of west.)

The farther the two east-west points are separated, the more you must angle northward to go from one to the other by the shortest route (or southward in Australia). It is even possible that if you place two points properly east and west, you will be forced to leave from one point in a due northerly direction to get to the other in minimum distance (and due southerly in Australia).

According to Euclid, the sum of the three angles of a triangle is 180°, but in a triangle drawn on the earth, ignoring any unevennesses in its surface, the sum of the angles is always *more* than 180° if we insist on sticking to the maximum-separation assumption. In fact, you can draw a triangle on the surface of the earth under conditions where each angle is a right angle, and the total sum is then 270°.

However, Hardhead has abandoned us long ago. Insanity is only funny up to a point and then it becomes infuriating.

Why set up the 20,000-kilometers separation maximum if it means that the consequences will violate straightforward geometry in so many ways? It may be the kind of game that could interest people who are fascinated with recreational mathematics, but would it not lead to a dangerous divorce from reality?

No! That's just what it does not do! When navigators make long voyages across the ocean, or airplanes fly long distances anywhere, or when you want to check the time with a friend in London, or do any of a number of things—you find that all the screwball consequences of your distance maximum must be taken into account. It is those consequences which actually describe the earth, and not Hardhead's "common sense."

Euclidean geometry is "plane geometry"; the geometry that is valid on a plane, by which is meant a perfectly flat surface. The surface of the earth, however, is not perfectly flat. It is easy to deduce from the assumption of a 20,000-kilometer maximum separation that the surface of the earth is spherical. The behavior of lines on its surface is described by the deductions of "spherical geometry" and everything I have mentioned is in accord with that, provided you consider the earth a sphere which is 6,370 kilometers in radius."*

But if the earth is a sphere and if all the rules of spherical geometry are well known, why didn't people understand the spherical nature of earth at once?

Because through most of history people were involved with very small patches of the earth's surface, across which the degree of curvature was vanishingly small. The surface was so close to flat that plane geometry was good enough, and since plane geometry is the simplest form of geometry, it came to seem "commonsense" and to represent universal truth.

To be sure, some Greek philosophers worked out the sphericity of the earth on theoretical principles, but it didn't really grab hold in general until the Age of Exploration began in the fifteenth century. Since it was then impossible to navigate successfully without taking earth's sphericity into consideration, the flat earth was discarded by all. (Well, there are a few amusing cranks who uphold it even today.)

Of course, that doesn't mean that everyone uses the maximum-distance assumption and understands its consequences. I simply chose that assumption because it can be made to sound ridiculous and yet will give the right answers.

Fortunately for us all, the spherical earth can be understood directly because it can be shown by a simple model. Paint the continents on a plastic sphere and you can quickly see what happens to Smith and Jones as they walk: their separations and approaches, the relationship of point Y to point X, the intersection of lines, the

* Actually, earth is not a perfect sphere but a slightly oblate spheroid, so the geometry is not quite as I described it, but the deviations are not significant.

reason for a northward or southward angle in traveling from east to west and so on.

The concept of a globe is so easily grasped that even old Hardhead capitulated.

But now let's try something else, by beginning with another aspect of common sense.

We all know that if we take a short run before trying a broad jump, we jump farther than if we tried it from a standing start. The speed of the run adds to the speed imparted to the jump by your thigh muscles, so that you start with a faster motion and go a greater distance before gravity pulls you to the ground again.

If you take careful measurements you will find that a ball thrown forward at a rate of 20 kilometers per hour (relative to the ground) will travel 40 kilometers per hour (relative to the ground) if thrown at its usual speed (relative to the thrower) while the thrower is traveling forward on a vehicle moving at 20 kilometers per hour.

On the other hand if the vehicle is moving at 20 kilometers per hour, and the thrower standing upon it throws a ball with a speed of 20 kilometers per hour in the direction opposite that in which the vehicle is moving, the ball travels 0 kilometers per hour relative to the ground and simply drops downward.

Furthermore, if two vehicles are approaching each other, each moving at 20 kilometers per hour relative to the ground, then a person on one vehicle will see the other approaching at 40 kilometers per hour relative to himself or herself.

To put it as briefly as possible, speeds add and subtract just as apples and oranges do, and since this is in accordance with Isaac Newton's laws of motion, you may think of it as part of a Newtonian universe. The Newtonian universe seems as commonsensical as Euclidean geometry, largely because it's about as simple as it can be.

Now let's pull an assumption out of left field, one that involves the addition and subtraction of speeds. Let us suppose that such manipulation of speeds does *not* work for anything moving at 300,000 kilometers per second.* Something moving at that speed relative to us does *not* change its speed relative to us when it is being carried forward or backward, in the direction of its travel or against it.

* Again, 299,792 kilometers per second would be better, but I'm using the convenient, and close, approximation.

Light, as it happens, moves at that speed when traveling through a vacuum so that when we measure the speed of light relative to ourselves, it always turns out to be 300,000 kilometers per second regardless of the motion of the source of light relative to us (or our motion relative to it).

What's more, anything that ordinarily moves at a speed less than light can't be made to move at the speed of light, let alone faster than the speed of light, because if it reaches the speed of light it will be trapped there, unable to move faster or slower. Similarly, anything that ordinarily moves faster than light (like the hypothesized tachyons) could never move as slowly as light, let alone slower.

In other words, any conceivable object in a vacuum travels either forever *at* the speed of light, forever *less than* the speed of light, or forever *more than* the speed of light. The speed of light is a barrier in both directions.

Why should that be so? What is so magic about that particular speed at which light moves?

No answer, really. That's just the way the universe is.

What are the consequences of that one assumption?

First, suppose two spaceships are moving away from each other, and each one is traveling 200,000 kilometers per second. Surely to each spaceship, the other spaceship seems to be receding at 200,000 + 200,000 = 400,000 kilometers per second?

No! The two figures have to be added in such a way that the key figure of 300,000 kilometers per second is not exceeded. A formula must be used which includes the ratio of the speed of the spaceship to the speed of light, and which will add any two figures, each below 300,000, in such a way that the sum is nearer 300,000 than either of the two figures being added, yet never quite reaches 300,000.

Again, imagine a spaceship flashing by you at enormous speed, and imagine further that you can observe and measure the time it takes light on the ship to travel from a source to a mirror and back. You will find that because the ship is moving so quickly, the light seems to be traveling a longer distance (relative to you) than it would if the ship were standing still (relative to you). Despite the fact that the light traveled a longer distance, the speed of light on the speeding ship is the same to you as it would be if the ship were standing still, for the speed of light in a vacuum never changes. Yet the light manages to cover the greater distance.

The only conclusion is that the rate of passage of time slows on a speeding ship. If time slows and a second grows longer, then light, without increasing its speed, can travel the greater distance.

In other words, rather than abandon our silly assumption that light never changes its speed, we have to assume that time slows with increasing speed.

Now common sense tells us that everything can change its speed if it's properly fooled with, while *nothing* can change the time rate, and therefore this tendency of relativists to do anything at all, even introduce the concept of variable time, just to save something as silly as the constancy of the speed of light, is simply enraging. The Hardheads can't endure it.

Nor is a variable time rate the only thing forced upon us by the constancy of light speed. In order to save that constancy, we have to have moving objects shorten in length in the direction of their motion as they speed up.

Then again, it turns out that the scheme of adding speeds in such a way that 300,000 kilometers per second is never exceeded makes an object more and more difficult to accelerate as that magic speed is approached. A force that is sufficient to increase its speed by 50 kilometers per second will, as the object approaches the speed of light, suffice to do so by only 20 kilometers per second, and, as it approaches the speed of light still closer, by only 5 kilometers per second, and so on. Finally, as an object moves infinitesimally close to the speed of light, all the force in the universe can only accelerate it infinitesimally.

This increasing difficulty of acceleration can be stated another way. We can say that a moving object increases its mass as it moves faster and faster, for the mass of an object is defined by the ease with which it accelerates.

The time rate, the length, and the mass of an object all vary with speed according to formulas including the ratio of the speed of the object to the speed of light. At ordinary speeds, the difference from the Newtonian situation is negligible (just as over small patches of earth's surface, the difference from flatness is negligible) while as the object approaches the speed of light, the time rate and length each approach zero and the mass approaches the infinite.

Imagine! All this screwiness just to save the constancy of the speed of light.

In fact, there's more. The energy of motion of a body—its "kinetic energy"—is measured as half its mass times the square of its speed.* This means that mass and speed are the only things involved in energy of motion. Ordinary speed, the addition of force in order to accelerate an object, adds to its kinetic energy by increasing its speed. It increases mass also but infinitesimally, so that the increase is never noticed at ordinary speeds and mass is assumed to be constant.

As the speed of an object gets progressively nearer the speed of light and the accelerating force has a smaller and smaller effect on the speed, it has a correspondingly greater and greater effect on the mass. More of the increase of kinetic energy is represented by the increase of mass. Thus, we have to accept the fact that energy can be converted into mass and, inevitably, vice versa. The relationship between the two is the famous $e = mc^2$, which is also necessary, then, to save the constancy of light speed.

Albert Einstein worked all this out in 1905 and went on to do much more in the succeeding decade. You make the one assumption and then have to alter a large number of other things, violating common sense to do so, just to keep the assumption going.

Is it worth it?

Yes! Physicists studying vast stretches of space, intense concentrations of energy, enormous speeds, find they cannot make head or tail out of what they observe unless they assume the correctness of the various equations of relativity. The thing is that light actually *has* a constant speed, and all the changes in length, mass, and time with motion, all the various relativistic modifications of the Newtonian universe, *really* exist. We live in an Einsteinian universe.

The Einsteinian universe seems against common sense only because all our ordinary experience of life deals with small regions and low speeds, where the relativistic corrections are virtually zero, and Newtonian relationships are correct to a high degree of accuracy.

Well, then, we've given up the flat earth and accepted the spherical earth. Why can't we give up the Newtonian universe and accept the Einsteinian universe?

Because there is no easily grasped model of the Einsteinian universe. We don't have the equivalent of a painted globe over which we can trace lines.

Suppose there were no such thing as a globe and no way of con-

* I should really be saying *velocity*. Speed and velocity are not quite the same thing in physics, but please excuse the imprecision this time because I'm making a conscious effort to use colloquial language in this essay in honor of the fact that I'm discussing relativity.

ceptualizing one. Suppose we had to work with a maximum separation as an assumption without ever explaining what this represented in the form of a spherical earth.

In that case, people would be demanding, to this day, *why* can't we go farther than 20,000 kilometers from home and yelling about it and getting red in the face with rage at scientists obtusely clinging to the limitation.

They would say, "If you were twenty thousand kilometers from home and kept on walking, you would somehow break the distance barrier. After all, we broke the sound barrier and we'll break the distance barrier, too. You scientists are just stupid and dogmatic."

But they *don't* say it because we can explain the situation with a globe and they see that the 20,000-kilometer distance is indeed a sensible maximum.

But in relativity, we have nothing easy to conceptualize: we must start with the constancy of light speed and deduce the consequences. And people can't accept it.

They say, "But *why* can't we go faster than light?"

And they say, "Suppose two spaceships are moving away from each other and each is going at two hundred thousand kilometers per second; doesn't it stand to reason that each spaceship sees the other as going faster than light?"

And they say, "We broke the sound barriers, and we'll break the light barrier. You scientists are just stupid and dogmatic."

And no amount of explaining ever seems to help.

13 * BEYOND EARTH'S EONS

When I was getting my education, I scorned commencements. I refused to attend them on the occasion when I earned my bachelor's degree, or my master's, but insisted the school mail me my diplomas. For my Ph.D., I broke down to the extent of attending the commencement, but I sat in the audience, refusing to go through the mummery of academic regalia.

I am well paid for that now. Since 1969, scarcely any commencement season has passed without my being forced to climb into cap and gown at least once. In 1976 I had to do it no fewer than four times. Almost always this is for the purpose of giving a commencement address and on a number of occasions for the purpose of garnering an honorary degree as well.

As a result I have collected, through 1979, four or five doctorates in science, one in engineering, one in letters, and one in humane letters.

Then came May 18, 1980, when I showed up at Boston University's commencement in order to collect another honorary degree. It was a happy occasion, for I am on the B.U. faculty as professor of biochemistry. I met some old friends and was made much of.

I did not know exactly which degree I would be awarded, but it didn't seem to make much difference. I had been granted one in every category for which there was the shadow of an excuse, so I expected nothing new.

Boston University fooled me. After some considerable discussion (I understand) as to whether it made more sense to give me a doctorate in letters or in science, president John Silber decided to do both and he made me a doctor of letters and science. Such a double degree was unprecedented for B.U. and (for all I know) may be unprecedented for higher institutions of learning generally.

I am not ordinarily sentimental about such things and have never framed any piece of academic parchment except for my Ph.D. diploma, but I am going to frame this new diploma and hang it on the wall.

And now that I have received this double honor, I will turn to a subject that is going to stretch my letters *and* science to the limit. I hope I make it.

In chapter 7, we discussed the age of the earth and came to a satisfactory figure of 4.6 eons, where we defined an eon as 1,000,000,000 (one billion) years. We pass on, now, to the question of the age of the universe generally.

Until the eighteenth century, there was no feeling anywhere that the two questions—the age of the earth and the age of the universe—were separate. It was always assumed, before the days of modern astronomy, that the heavenly bodies were a minor adjunct of the earth. The Bible says, "In the beginning God created the heaven and the earth," and if other religions and philosophies didn't use those words, they had the concept—that everything was created at once.

In the eighteenth century, when scientists began to speculate about a nondivine creation of the solar system, either through some catastrophic event involving the sun or through the evolutionary condensation of a vast mass of dust and gas, it seemed reasonable to suppose that this was a local phenomenon. The sun, in the catastrophic case, might well have existed for an indefinite period before the birth of the earth, along with its millions of companion stars. The mass of dust and gas, in the evolutionary case, might well have existed for an indefinite period before the birth of the solar system, and other stars and their planetary systems might already have formed eons before.

In either case, the universe was surely older than the solar system, possibly very much older. In fact, if one could get away from the notion of a divine creation (not an easy thing to do in the eighteenth and nineteenth centuries), the universe might even be infinitely old.

In chapter 7, I pointed out that the advancement of the law of conservation of energy in the 1840s made it clear that the sun must have begun with some fixed store of energy which would someday be used up. In short, as an energy-radiating body, it had to have a beginning and an end.

The law of conservation of energy did not, in itself, imply the same for the universe as a whole. Individual stars might eventually consume their energy source, but new stars might form, and this might continue indefinitely.

Why not? The law of conservation of energy (also called the *first law of thermodynamics*) held that energy could never be created or destroyed, but could only be transferred from one place to another or changed from one form to another. The sun, in using up its energy, merely transferred that energy from itself into surrounding space, and other stars did the same. Might not the energy flooding into space come together to form as many new stars as had previously died and continued to do so forever?

This dream was ended by a German physicist, Rudolf J. E. Clausius (1822–88). In 1850, he discovered that if one considered the ratio of the heat content to the absolute temperature of a closed system (one that lost no energy to the outside world and gained no energy from it), that ratio always increased in the course of any spontaneous change taking place within that closed system.

Clausius named the ratio *entropy,* so one could say that in any closed system, all spontaneous changes involved an increase of entropy and this came to be called the *second law of thermodynamics*.

The universe as a whole is, as far as we know, a closed system—

in fact, the only truly closed system. Therefore we can say that the entropy of the universe is constantly increasing.

We don't know, offhand, how high the increase can go, but suppose we look at matters in reverse. If the entropy of the universe is constantly increasing, the total entropy of the universe was less last year than it is now and still less the year before and even less the year before that.

If we assume that the total entropy right now is finite, then if we move back in time far enough, we will find ourselves with an entropy of zero. It would seem that we can't move backward further than that, so that if the second law of thermodynamics is true (and every observation we have made in over a century and a quarter since Clausius' time leads us to believe it is), then the universe had to have a beginning.

It turns out that entropy is a measure of the unavailability of energy. Energy can be turned from one form into another, but not with perfect efficiency. Every transformation leaves less of the total energy available for conversion into work. While the total energy of the universe remains constant with time, less and less of it is available for conversion into work as time goes on.

Eventually, all the energy, if we assume a finite amount, becomes unavailable. At that time, all the energy is in the form of heat spread out evenly with no temperature differences. Under such conditions, entropy is at a maximum, and available energy is zero. While we can still speak of a universe, for the energy is still all there, it is a universe without any further possibility of change—life—us. For all practical purposes, the universe is dead and, indeed, Clausius spoke of "the heat death of the universe."

To summarize: Given the second law of thermodynamics and a finite universe, we can deduce from that alone that the universe must have had a beginning and must someday have an ending.

We cannot, from the second law of thermodynamics alone, decide *when* the beginning was, or *when* the ending will be. That depends on the total energy content of the universe and the rate at which entropy is increasing, and Clausius couldn't even begin to guess at those figures.

Of course, we don't have to know all the details of energy content and rate of entropy increase. Suppose we find some change taking place in the universe, a change so vast, so steady, so unidirectional, that we can assume the entropy increase with respect to that change utterly swamps all other, lesser entropy increases. We can then pretend that that one change is all that is taking place and work with that. Everything else merely introduces insignificant modifications that don't perceptibly affect the final answer.

The chance of finding such a change would surely have seemed tremendously small and yet it is there and it was found.

The possibility of the discovery dates back to the 1840s, when it was shown that from the radiation given off by a moving object one could tell whether that object was approaching us or receding from us and at what velocity it was doing so.

Beginning in 1912, the technique was applied to the light spectra of certain "nebulas" in the sky which were thought to be clouds of dust and gas not terribly far away. They turned out, however, to be very distant objects.

The stars visible to us with the unaided eye are part of an enormous structure called the galaxy, made up of several hundred billion stars, but beyond the galaxy are other galaxies, some as enormous as our own and some more enormous still. These outer galaxies stretch through vast distances and the "nebulas" proved to be these distant galaxies.

The American astronomer Edwin Powell Hubble (1889–1953) first clearly demonstrated this when, in 1917, he used the then-new 100-inch telescope at Mt. Wilson in California to take photographs of the rim of the Andromeda "nebula" (the only one visible to the unaided eye) and found the cloudy luminosity to be the result of enormous numbers of very faint stars.

By the time Hubble accomplished this task, it had been shown that all the galaxies studied (except for one or two of the nearest) were receding from us and some were doing so at surprisingly large velocities.

Hubble grew interested in this. He collected all the data available on the speed of recession of the various galaxies and pressed for more and more observations of this sort. He correlated the speed of recessions with the relative distance of the galaxies (using various methods to determine those relative distances) and it became plain that there was a simple linear relationship. The farther a galaxy was, the faster it receded from us. If galaxy 1 were five times as far from us as galaxy 2, then galaxy 1 was receding at five times the velocity that galaxy 2 did.

By 1929, Hubble felt it safe to announce the relationship, which has been called Hubble's law ever since.

It may seem very peculiar to have all the galaxies receding from us as though they were repelled by us, with the force of repulsion increasing with distance. Actually, though, that is the wrong way to look at it. A more sensible interpretation of Hubble's law is to suppose the entire universe is expanding. If we suppose this, then, from the viewpoint of an observer on *any* galaxy and not only our own, the distant galaxies would be receding at a rate proportional to their distance.

Such an expanding universe is consistent with the equations worked out by Albert Einstein (1879–1955) in 1916 in his general theory of relativity.

The vast expansion of the universe is so enormous a phenomenon that we can work with it alone to consider the beginning and ending of the universe, and assume that it will give us the same answer we would have gotten by working with the second law of thermodynamics in full detail.

For instance, the steady expansion of the universe means that last year it was smaller than this year, and the year before smaller still and so on, until at some time in the long distant past it was no larger than a point. It was at that past moment of time that entropy was zero. The universe began at that moment in a "big bang" which resulted in an explosive expansion.

Let us pretend now that we are dealing with a galaxy that is 13 million light-years from us and that is receding from us at a speed of 2,000 kilometers per second. If we imagine time moving backward, then the galaxy is approaching us and, every second, is 2,000 kilometers closer to us than it was before.

Since there are 31,557,000 seconds in a year, that means that the galaxy is roughly 63 billion kilometers closer to us every year we move backward in time. A light-year is equal to 9,460 billion kilometers, so that we would have to move backward in time some 150 years before the galaxy would shave a single light-year off its distance.

To have the galaxy move 13 million light-years we would have to move back in time 150 × 13,000,000, or roughly 2 billion years; that is, two eons. In other words, two eons ago, that galaxy and ours occupied the same place.

What holds for that galaxy would hold for any galaxy if Hubble's law is correct. If galaxy 1 is twice as far as galaxy 2, galaxy 1 is moving twice as fast so that both will cover the distance by which they are separated from us in the same time. You can argue it out similarly for all the galaxies.

In short, if Hubble's law holds, then all the galaxies will coalesce at the same time if we imagine time moving backward. To calculate that time it is only necessary to choose one distant galaxy and work out both its distance and its velocity of recession. Divide the former by the latter and you will find the time that has elapsed since the big bang and thus you will have determined the age of the universe.

There's no doubt about the velocity of any galaxy whose spectrum can be studied. How the velocity affects the dark spectral lines is completely understood, and it can be checked by observations in the laboratory.

That leaves distance, and there, unfortunately, we are dealing with something not at all easy to determine. When the twentieth century began it was not possible to determine distances of more than 100 light-years with any degree of reliability.

In 1912 it was discovered, however, that certain variable stars called *Cepheids* had periods that varied with their actual luminosity. The period of light fluctuation was easily measured and from that the luminosity of each Cepheid could be calculated. If the apparent brightness is compared to the luminosity, the difference must be due to the light-diluting effect of distance. In this way, the distance of that Cepheid or of any structure of which the Cepheid forms a part can be determined.

When Hubble first worked out his law, he determined the distances of some of the nearer galaxies by detecting Cepheids in their outskirts. Those Cepheids were much fainter in appearance than Cepheids of the same period in our own galaxy. Since they had the same period they must actually be of equal luminosity. The extreme faintness of the Cepheids in the outer galaxies must be the result of their great distance and Hubble worked out those distances.

Once he had done so, he divided distance by velocity and decided (as I did a little earlier in the essay) that the universe was two eons old.

This was a horrible shock for geologists and physicists. Judging from the study of uranium and lead in the rocks (as I explained in chapter 7), they were convinced that the earth was considerably older than two eons. Surely, it was inconceivable that the universe was younger than the earth.

In 1942, fortunately, the German-American astronomer Walter Baade (1893–1960) found reason to believe one should divide stars into two classes: population I and population II. He was able to show that for a given period of light fluctuation, Cepheids varied in luminosity depending on whether they were population I or population II, with the former being considerably more luminous for a given period.

In studying the outer galaxies, Hubble had been observing population I Cepheids and had applied to them the rules for population II Cepheids. Thinking those distant Cepheids were less luminous than they really were, he ended with moderate distances for the galaxies.

Once the Cepheids were recognized for the much more luminous population I variety, the galaxies were seen to be far more distant than Hubble had thought, if that vast luminosity were to be reduced to the observed tiny spark.

Applying Baade's insight, it was seen that the greater distances divided by the same old velocity (which there was no reason to change) gave a correspondingly greater quotient so that the time of the big bang had to be placed farther into the past and the universe was seen to be far older than two eons and, indeed, far older than the 4.6-eon age of the earth. Geologists and physicists heaved sighs of relief.

For thirty years after Baade's work, every time astronomers discovered new ways of determining the distances of the galaxies, the figures seemed to be larger than had previously been thought and the universe, therefore, older.

By 1979, the distances were roughly ten times what Hubble had thought at first and the age of the universe was thought to be (by a straightforward application of Hubble's law) twenty eons old.

A straightforward application, however, is oversimple.

The universe is expanding against the gravitational pull of all its parts. The task of overcoming that pull deprives the galaxies of kinetic energy and, in their expansive outward rush, they move more and more slowly with time (just as a ball moving upward against the pull of earth's gravity moves upward more and more slowly with time).

This means that if we imagine ourselves going backward in time, we will see the galaxies approaching each other more and more rapidly and they will coalesce in less time than the twenty eons arrived at by supposing their speed of recession is constant throughout.

How much less than twenty eons the age of the universe is depends on how strong the universe's gravitational pull is. That, in turn, depends on what the average density of the matter of the universe is. The denser the universe is on the average, the stronger its gravitational pull, the shorter the time since the big bang and the younger the universe.

Unfortunately, we're not sure about the density of the universe so we can't make any hard-and-fast deductions. All we can say is that twenty eons would seem to be the maximum age of the universe, and if it is, on the whole, not very dense, then the actual age may not be much below that.

In that case, the solar system is just about one-fourth as old as the universe is. To put it another way, the universe existed (and did very well, I'm sure) for three times as long without the sun and the earth as it did with them.

Once we determine the age of the universe, we have automatically determined the size of the universe. At the moment of the big bang, electromagnetic radiation such as light began to speed outward in all directions—at the speed of light, of course.

The expanding globe of radiation moved outward one light-year for each year of time that passed and one light-eon (a billion light-years) for each eon of time that passed. When Hubble thought the universe was two eons old, it would have had to be 2 billion light-years in radius and, of course, 4 billion light-years in diameter.

As astronomers' estimates of the age of the universe increased, so did its estimated size. If the age is twenty eons, the radius is 20 billion light-years and the diameter 40 billion.

The figures for the age and size of the universe are by no means hard and fast, however. The measurements used are highly ingenious but they are pretty close to the borderline in accuracy. Instruments must be used close to their limits and reasoning must rest on possibly shaky assumptions. It would not therefore be surprising if further observations resulted in further alterations of the distance to the far galaxies—and therefore of the age and size of the universe.

In late 1979, three American astronomers, Marc Aaronson, John Huchra, and Jeremy Mould made use of some new techniques for determining galactic distances.

For one thing, they studied the globular clusters which are associated with galaxies, including our own. These clusters are comparatively small, comparatively dense spherical accumulations of anywhere from 10,000 to 1 million stars. Each galaxy possesses one or two hundred of them in a wide range of luminosity.

You can't tell much by comparing one distant galactic globular cluster with another that is attached to a different galaxy, because the two may be of different sizes. The three astronomers, however, noted that in various galaxies of the Virgo galactic cluster, the *range* of brightness was about the same for all.

It might be that the range was always the same for all galaxies, including our own. By comparing the apparent brightness of the range of globular clusters of a distant galaxy with that of the range of globular clusters of our own galaxy, we can calculate the distance required to reduce the brightness of the former to the dim sparks actually observed.

Then again, the three astronomers measured the rate at which distant galaxies were rotating. This can be done if the galaxies happen to be viewed edge-on from earth. If the spectrum is taken first at one end, then the other, there will be a red shift in one case and a blue shift in the other and from the size of those shifts the rotation period can be determined.

The faster a galaxy is rotating at its edges, the more massive it must be, for it is the mass that produces the gravitational field that whips along the movement of the stars. Once the mass is determined, the actual luminosity is also known. If this luminosity is compared with the actual brightness we see, an estimate of the distance of the galaxy can be made. What's more, the three astronomers measured the brightness in the infrared, which is much less likely to be scattered by dust and thus dimmed across the vast distances than visible light is.

Using these methods, the three astronomers presented evidence to the effect that previous distance figures rather overstated the case and that actually, the distances of the galaxies ought to be cut in half.

If they are correct, it means that the universe is only ten eons old and is only 20 billion light-years in diameter. It would also mean that the solar system was about half the age of the universe.

When this was first announced, there was considerable excitement among some nonastronomers. At least I got calls from reporters whose questions to me made it seem they thought that the world of science had been turned upside down and that astronomy had been exposed as a rickety science because "all of a sudden half the universe had disappeared."

The things to remember in this connection are:

1. We are dealing, as I have explained, with very borderline measurements.

2. The universe has, in thirty years, been expanded tenfold in age, from two eons to twenty eons, as astronomers sharpened their observations so that being pushed back to ten eons (a twofold reduction) is a comparatively minor adjustment.

3. The returns aren't all in yet. The longer age seemed to fit not only the distance measurements but the time required for certain facets of stellar evolution. The reduced figure of ten eons may not leave quite enough time for these facets—which makes many astronomers reluctant to accept the new figure.

So we'll see. Undoubtedly many astronomers are now checking the new work and are going over the older work, and we shall continue to sharpen the manner in which the eons are being counted, to the good of everyone.

14 ★ THE CRUCIAL ASYMMETRY

On rare occasions I sit in a bar as a matter of social necessity and a few nights ago such a necessity arose. I ordered my inevitable ginger ale and observed with scientific detachment (well, not quite) the beautiful barmaid, whose long legs were covered by nothing more than sheer hose for their full length.

I was moved to philosophical reflection and said, out loud, what I have often thought.

"The rewards in male-female relationships," I said, "seem to be weighted enormously in favor of the male. Consider the female leg —how utterly smooth, graceful, well proportioned, and (I happen to know) delightful to the touch. What do women get in return for what they have to offer? The male leg: hairy, bunchy, and (I suspect) equally repulsive to sight and touch."

Whereupon a young lady, who was also at the table, said in wide-eyed wonder, "How can you possibly manage to get the situation so completely reversed?"

That left me speechless, I assure you, but as I lingered over my ginger ale, I worked it out in my mind. The natural attachment of men for the consummate charms of women, as contrasted with the bizarre affection women feel for men, is a necessary and even crucial asymmetry that preserves the human species.

It's tough on women, of course, but apparently that is how it must be.

And having straightened that out, let us move on to a crucial asymmetry in the universe that is even more broadly significant.

The universe is thought to be about 15 billion years old in the sense that 15 billion years ago there was a "big bang." With the big bang, the universe came into existence as an object with its present mass but with a diameter that was virtually zero and a temperature that was virtually infinite.

With incredible rapidity, it expanded and cooled, continuing to do so at a steadily slowing rate. It continues to expand and cool today, 15 billion years later.

At the start, with mass and energy unbelievably concentrated, changes naturally took place very rapidly. They had to. For one thing, all change is driven by energy and there never was such a concentrated energy supply in our universe as existed at the very start. Secondly, change is made easier and more rapid if the constituent bits that are being subjected to change are close together so that they can interact without undue delay, and there never was such closeness in our universe as existed at the very start.

As the universe expanded, its constituent bits spread farther apart and the energy concentration (temperature) declined. For both reasons, the rate of change in the universe slowed with expansion.

Because of the enormous rate of change at the start, physicists talk about crucial events that happened only minutes after the beginning, and only seconds after, and only very tiny fractions of a second after. Carefully, they calculate what must have happened in less than a billionth of a billionth of a second after the big bang.

It shakes the mind. How can so much happen in such an ultrabrief interval?

Ah, but let's not consider time as a smoothly and evenly flowing stream, with every second just as filled with potential events as every other second. Let us not consider all seconds as tiny bags of events of precisely the same size.

We are lured into thinking of seconds as equally sized and equally eventful right now, because the expansion-cooling of the universe is now proceeding at a rate so small compared to its present size and temperature that there is no perceptible change in the number of events a second can hold (on the average over the universe generally) in a human lifetime or even in recorded history. There's not much change even over stretches of millions of years.

In the early beginnings of the universe, however, seconds were incredibly crammed with events. An early second could hold trillions of times as many events as a contemporary second. A still earlier second could hold trillions of trillions of times as many events as a contemporary second. Judging by how many events they could hold, an early second was the equivalent of thousands of contemporary years in length, while a still earlier second was the equivalent of millions of contemporary years in length, and so on.

If we measure time by events, it would make more sense to treat time logarithmically. Let us not suppose it behaves arithmetically, so that one-fifteenth of all the events that have ever taken place took place in the first billion years; another one-fifteenth in the second billion years; another one-fifteenth in the third billion years; and so on, until the final one-fifteenth took place in the fifteen billionth year, which brings us to now.

Let us suppose, instead, that half of all the events that took place in the universe took place in the first one-tenth of the universe's lifetime; that half of that half took place in the first one-hundredth of the universe's lifetime; that half of that half of that half took place in the first $\frac{1}{1000}$ of the universe's lifetime and so on. This is the logarithmic view.*

This means that half the events that have ever taken place in the universe had taken place by 1,500,000,000 years after the big bang; a quarter of the events by 150,000,000 years after the big bang; an eighth of the events by 15,000,000 years after the big bang, and so on.

But let's not work with all those zeroes. Let's use exponential figures instead. In place of 15,000,000,000 years, we can say 1.5×10^{10} years, where 10^{10} is the product of ten tens multiplied together, or a 1 followed by 10 zeroes. Again, 1,500,000,000 years is 1.5×10^9; 150,000,000 years is 1.5×10^8; 15,000,000 years is 1.5×10^7; and so on. Now, we have exponents that go down smoothly from 10 to 9 to 8 and so on and since exponents are very closely related to logarithms, this gives us a logarithmic scale.

For convenience, in fact, let us consider the age of the universe in seconds. Each year is made up of 31,556,926 seconds so that

* Matching events by halves to durations by tenths is just a matter of convenience in calculation. The actual match may be different, and more complicated, but I am just me and not a theoretical physicist.

15,000,000,000 years is just about 470,000,000,000,000,000, or 470 quadrillion, seconds long. Exponentially, we can write it as 4.7×10^{17} seconds.

Let us set up a logarithmic scale of time by drawing a straight line divided into equal intervals marked 1, 2, 3, and so on up to 18, as in figure 1 on the next page.

Point 1 would represent the time 10^1, or 10 seconds, after the big bang; point 2 would be 10^2, or 100 seconds, after the big bang; point 3 would be 10^3, or 1,000 seconds, after the big bang; and so on up to point 18, which would be 10^{18} or 1,000,000,000,000,000,000 seconds after the big bang. The universe at present is located at about point 17⅔.

Each time interval on the line would seem to be ten times as long as the one immediately above it by ordinary arithmetic. Thus, the interval between 2 and 3 is the interval from 10 to 100, or 90 seconds; while the interval between 1 and 2 is the interval from 1 to 10, or 9 seconds.

In terms of the number of events that took place within them, however, each interval may be considered as long as every other. As many events took place between 1 second and 10 seconds after the big bang as took place in all the billions of years representing the stretch between points 17 and 18 (100,000,000,000,000,000 seconds and 1,000,000,000,000,000,000 seconds).

Yet need we begin the logarithmic line of figure 1 at 1? Might we not extend it further upward to 0?

Certainly we can—and the first thought might be that 0 represents the big bang itself, but it doesn't!

The 0 on the line is not an ordinary zero but an exponential zero, representing 10^0 seconds, and 10^0 is *not* zero, though that might seem logical at a quick glance. Since 10^2 (100) is $\frac{1}{10}$ of 10^3 (1,000), and 10^1 (10) is $\frac{1}{10}$ of 10^2 (100), then it is reasonable to suppose that 10^0 is $\frac{1}{10}$ of 10^1 (10). But $\frac{1}{10}$ of 10 is 1; therefore 10^0 should be equal to 1, and 0 on the logarithmic scale should represent 1 second.

We can extend the scale still further upward, and have —1, —2, —3, and so on, with each figure representing a point ten times closer to the big bang than the one before. Thus —1 represents the point one-tenth of a second after the big bang, —2 the point one-hundredth of a second after the big bang, —3 the point one-thousandth of a second after the big bang and so on (see figure 2).

Figure 1. Logarithmic Time

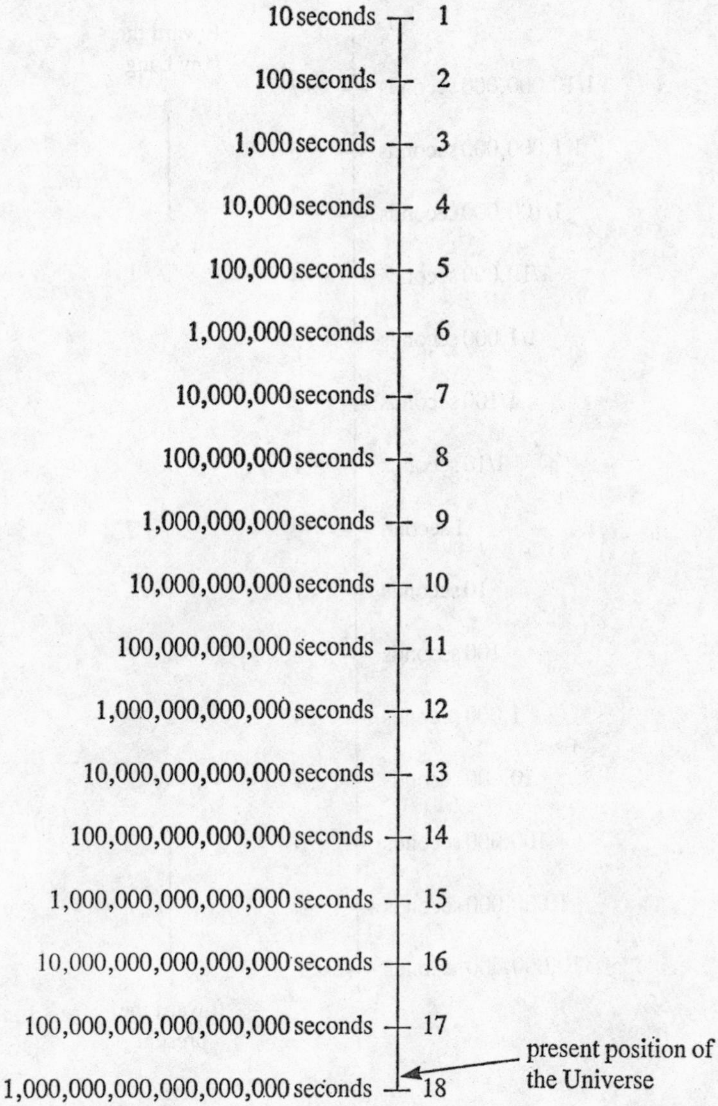

10 seconds	1
100 seconds	2
1,000 seconds	3
10,000 seconds	4
100,000 seconds	5
1,000,000 seconds	6
10,000,000 seconds	7
100,000,000 seconds	8
1,000,000,000 seconds	9
10,000,000,000 seconds	10
100,000,000,000 seconds	11
1,000,000,000,000 seconds	12
10,000,000,000,000 seconds	13
100,000,000,000,000 seconds	14
1,000,000,000,000,000 seconds	15
10,000,000,000,000,000 seconds	16
100,000,000,000,000,000 seconds	17
1,000,000,000,000,000,000 seconds	18

present position of the Universe

Figure 2. Extended Logarithmic Time

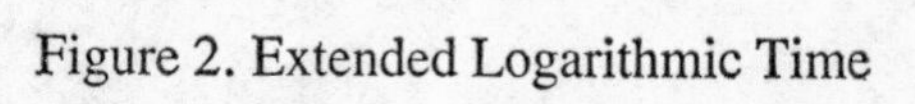

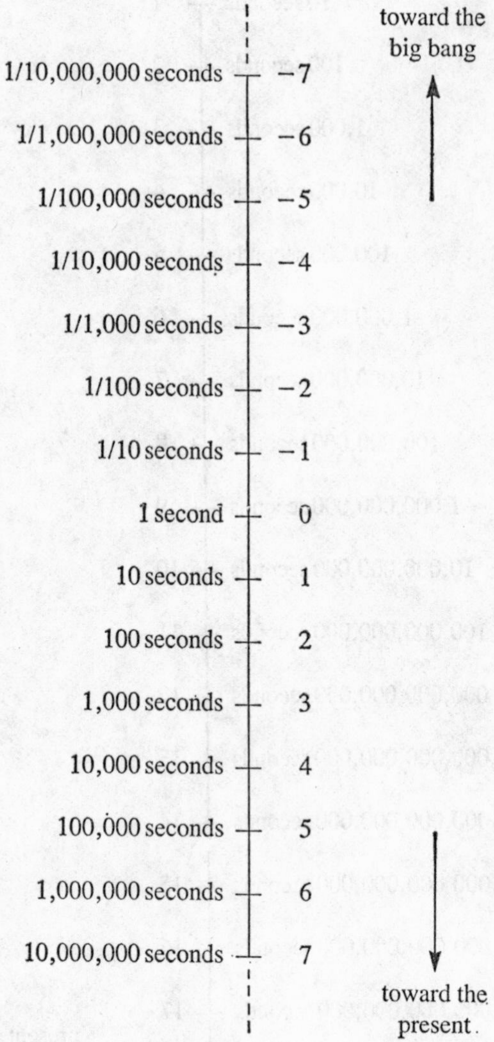

From the standpoint of the number of events taking place, each of these new tiny intervals is as long as any of the others. As many events took place between a thousandth of a second and a hundredth of a second, or between a thousand-millionth of a second and a hundred-millionth of a second after the big bang, as took place in the last 13.5 billion years.

If we imagine ourselves moving along the logarithmic time line toward larger and larger numbers, the universe is expanding and cooling. At 17⅔, where the universe is now located, it is about 25,000,000,000 light-years in diameter and has an average temperature of 3° K.; that is, 3 degrees above absolute zero. (Of course, there are places where the temperature is much higher than that in today's universe, even very much higher, as at the center of the sun, but we're talking average.)

If we imagine ourselves moving along the logarithmic time line toward smaller and smaller numbers, the universe is contracting and warming. (From our present taken-for-granted view, it would be like taking a film of the universe as it expands and cools, and running it backward.)

Suppose, then, we run the film backward until the universe reaches point 12. At that point, the universe is 10^{12} seconds (or about 100,000 years) after the big bang. When the universe is that young it has had time to expand to a diameter of only about 200,000 light-years. Its total width is not very much greater than the total width of a large galaxy of the contemporary universe. The average temperature of this small universe is estimated to be about 1,000° K.

If we imagine the universe to have contracted to this size, there is no longer room for galaxies, or even stars. The universe is a mere chaos of atoms. In fact, if we try to trace the universe back to points smaller than 12, it becomes too hot and too crowded for matter to exist even as atoms.

In short, it is only 100,000 years after the big bang that atoms formed and ordinary matter began to exist. It was only after a period of further expansion and cooling that the atoms could gather into stars and galaxies. Those atoms, stars, and galaxies making up the matter universe have existed for almost all the time of existence (considered arithmetically) of the universe as a whole—14.9 out of the 15 billion years. All those billions of years, however, take up only the last 5⅔ intervals of the logarithmic time line.

What existed at points lower than 12? What existed before the matter universe?

When the universe was too small and too hot to contain atoms, it must have been made up of the various subatomic constituents of atoms. It must have been a mélange of photons, electrons, neutrons, protons, and so on. Of these the neutrons and protons make up the nuclei of normal atoms and are lumped together as *nucleons*. They are the most massive particles at this stage and we can refer to the nucleon universe as existing at points less than 12 on our logarithmic time line.

If we imagine the universe moving through points smaller and smaller than 12, the nucleons are pushed closer and closer together and the temperature gets higher and higher. By the time we reach —4 on the scale, the universe is only about 250 kilometers (155 miles) across, no larger than an asteroid, though it still contains all the mass of the present-day universe. The asteroid-sized universe has an average temperature of about 1,000,000,000,000° K. (a trillion degrees).

At points smaller than —4, the universe is too small and too hot for nucleons to exist. In other words, nucleons did not form until $\frac{1}{10,000}$ second after the big bang. From $\frac{1}{10,000}$ second to 100,000 years after the big bang we had the nucleon universe and from 100,000 years after the big bang to the present, we have had the matter universe.

In point of ordinary arithmetical time, the matter universe has lasted 150,000 times as long as the nucleon universe. What's more, the matter universe is steadily, if slowly, increasing the ratio of endurance as it continues to exist for what will undoubtedly prove to be many billions of additional years.

On the logarithmic scale, however, which measures time by events rather than by arithmetical duration, the nucleon universe has lasted for 16 intervals (—4 to 12) while the matter universe has so far lasted only 5⅔ intervals (see figure 3). This means that about three times as many events took place during the apparently short lifetime of the nucleon universe as during the apparently long lifetime of the matter universe. The matter universe would have to continue to endure for something like 10,000,000,000,000,000,000,000 additional years in something like its present form in order to equal the nucleon universe in event content.

Until recently, physicists could only go back to about —4 on the logarithmic scale. They could not tell what was taking place within

Figure 3. Nucleon-Universe

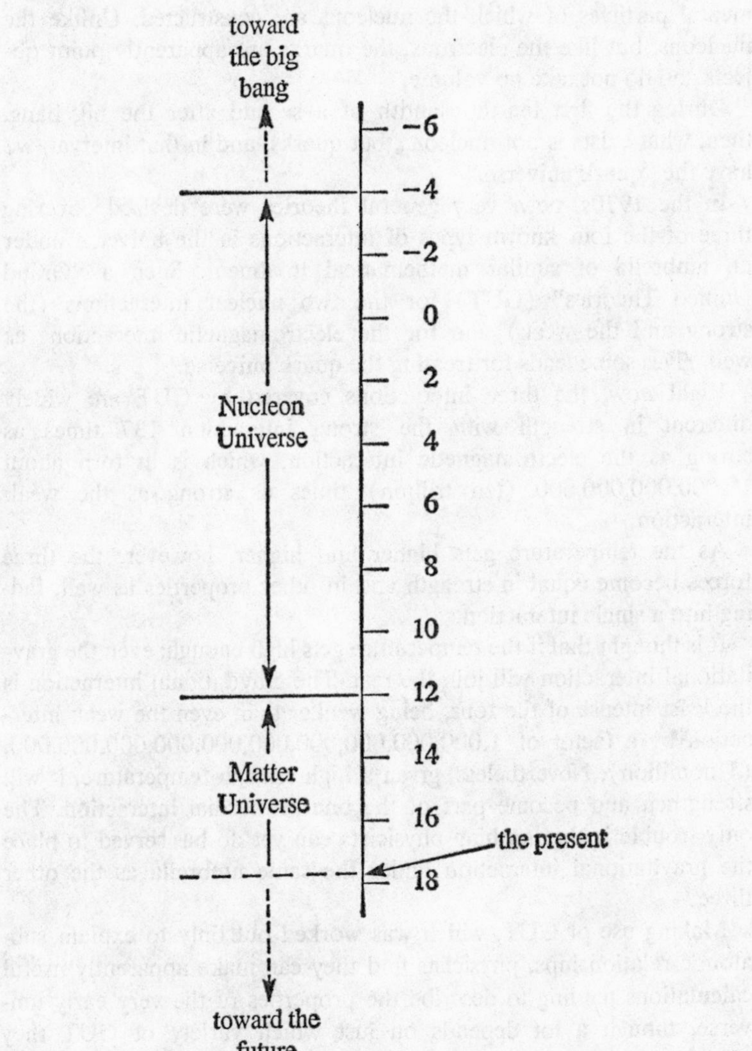

the first $1/_{10,000}$ second after the big bang, when the universe was far smaller than a sizable asteroid and far hotter than a trillion degrees. All they could say was that nucleons could not exist.

The concept of quarks arose in the 1960s. They are the fundamental particles of which the nucleons are constructed. Unlike the nucleons, but like the electrons, the quarks are apparently point objects and do not take up volume.

During the first ten-thousandth of a second after the big bang, then, what exists is not nucleons, but quarks, and in that interval, we have the "quark universe."

In the 1970s, new, very general theories were devised covering three of the four known types of interactions in the universe under an umbrella of similar mathematical treatment. Such a "Grand Unified Theories" (GUT) for the two nuclear interactions (the strong and the weak) and for the electromagnetic interactions as well, gives some leads for treating the quark universe.

Right now, the three interactions covered by GUT are widely different in strength with the strong interaction 137 times as strong as the electromagnetic interaction, which is in turn about 10,000,000,000,000 (ten trillion) times as strong as the weak interaction.

As the temperature gets higher and higher, however, the three forces become equal in strength and in other properties as well, fading into a single interaction.

It is thought that if the temperature gets high enough, even the gravitational interaction will join the rest. The gravitational interaction is the least intense of the four, being weaker than even the weak interaction by a factor of 1,000,000,000,000,000,000,000,000,000,000, (1 nonillion). Nevertheless, given a high enough temperature, it will strengthen and become part of the one primordial interaction. The only trouble is that nothing physicists can yet do has served to place the gravitational interaction under the same umbrella as the other three.

Making use of GUT, which was worked out only to explain subatomic relationships, physicists find they can make apparently useful calculations tending to describe the properties of the very early universe, though a lot depends on just which variety of GUT they use.*

* Which variety will win out depends on the exact results of delicate measurements of phenomena such as proton decay, which physicists hope eventually to be able to make.

Right now, some physicists trace the universe all the way back to point —43 on the logarithmic scale. That point is $1/10^{43}$ seconds (one hundred duodecillionths of a second) after the big bang.

At 10^{-43} seconds, the universe is of infinitesimal size compared even to a proton, for instance, and has an average temperature of 100,000,000,000,000,000,000,000,000,000,000° K. (100 nonillion degrees). At such a temperature, gravitational effects are strong enough to have to be taken into account, but the theory to do so is lacking. Physicists, therefore, cannot go to numbers smaller than point —43. The quark universe goes from —43 to —4. Points smaller than —43 represent an unknown region (see figure 4).

As you see, the quark universe lasted 39 divisions on the scale as compared to 16 divisions for the nucleon universe and 5⅔ for the matter universe. And who knows how many divisions the prequark universe lasted or how many different types of universe there were in the crucial duodecillionths of a second before quarks came into existence?

In fact, if we insist on looking backward in time through a logarithmic scale, it would seem that the universe extends an infinite distance through a never-ending series of intervals, each equal in amount of change content. To the left of —43 are —44, —45, and so on indefinitely. Can it be that we must conclude the big bang never really took place because it took place an infinite number of divisions ago containing an infinite number of events?

I don't believe that. Here's my own *guess* about it.*

As one moves farther and farther down the scale into smaller and smaller numbers, and the universe becomes ever more compressed and ever hotter, there is a tendency to form more and more massive particles. These ultramassive particles could not possibly exist today because there is nowhere energy sufficiently concentrated to form them. Such an energy concentration did exist in the very early universe, however.

In that case, if we go far enough back in time, might not the universe have been tiny enough and hot enough for a *single particle* to exist with the mass of the entire universe? My own selection for the name of such a particle would be a *holon* (the "particle of the whole," so to speak).

Perhaps that is the ultimate retreat pastward on the logarithmic scale. The big bang consisted, we might speculate, in the creation of

* Most of the rest of this chapter consists of my own guesses. No one else is to be blamed for it.

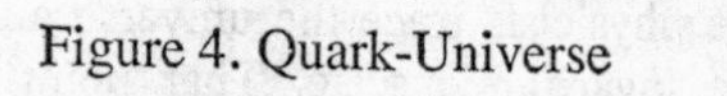

Figure 4. Quark-Universe

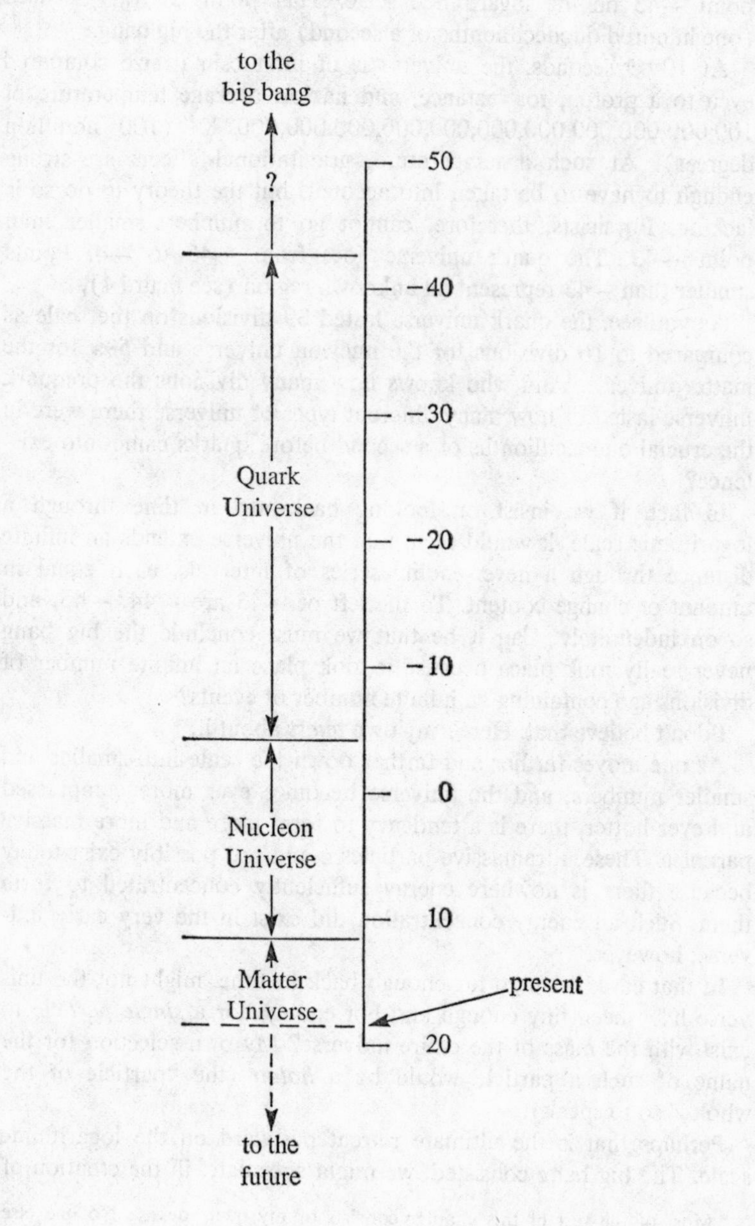

a holon with the mass of a hundred billion galaxies squeezed into a single particle with an unimaginably small volume; a particle as small relative to a proton, perhaps, as a proton is to the present-day universe.

Suppose that if we go backward along the logarithmic scale into the far, far past, the compression and temperature at point —100 is enough to make it possible for the holon to exist. Why not consider that the big bang takes place at —100?

The holon is unstable and breaks down into smaller particles in an unimaginably brief moment of time. These smaller particles repel each other intensely so that the universe spreads outward and, in consequence, cools. The smaller particles break down further in a slightly longer moment of time and so on. The expansion and cooling continues and at —43, quarks are beginning to form and the single existing interaction divides into two forms, one of which rapidly weakens with falling temperature to become what we now recognize as the gravitational interaction.

As expansion-cooling continues, two other interactions break away and weaken, taking us on the high road to the present-day four interactions which, in the low temperatures of the contemporary universe, seem irretrievably different.

Meanwhile, at —4, the quarks combine with each other to form nucleons, and at 12, the nucleons combine with each other or with electrons (or both) to form atoms and matter. As we progress past 12, the chaotic mélange of atoms breaks up into galaxy-cluster-sized turbulences and these condense into stars so that the universe becomes the familiar one of today—the slow-changing, incredibly slow-motion appendix to the long, exciting, and immensely changeful universes of the far past, all of which were over (by arithmetic time) in less time than *Homo sapiens* has existed.

At —4, where prequark particles form quarks for the first time, we are at an important crossroad, for here we encounter something we call the *law of conservation of baryon number*. This law, deduced from observations in our contemporary universe, states that the total number of baryons minus antibaryons in the universe must remain constant. (An antibaryon is precisely the same as a corresponding baryon in almost all properties except a particularly important one, such as electric charge or magnetic orientation, in which it is precisely opposite.)

If a baryon is destroyed, a balancing antibaryon must be destroyed with it, and both converted into photons. If a baryon is brought into existence out of photons, then a balancing antibaryon must also be brought into existence. In either case, the total number of baryons minus antibaryons remains unchanged. Included among the baryons are the protons and neutrons, as well as the quarks that make them up.

Thus, a proton and antiproton can undergo mutual annihilation, as can a neutron and antineutron. In addition, a quark and antiquark of the same color and flavor can annihilate each other. Again protons and antiprotons must be produced in pairs, as is true of neutrons and antineutrons, and (given equivalence in flavor and color) quarks and antiquarks.

Consequently, when prequark particles break down, it would seem that they must form quarks and antiquarks in equal numbers. As the temperature drops further, those quarks and antiquarks would annihilate each other to produce photons. The universe would end up with no nucleons at all, and therefore no atoms, and therefore no stars and galaxies. In short, the universe, as we know it, would not exist.

Where, then, did the universe come from?

GUT makes it plain, however, that the theory of conservation of baryon number is not true, but only *almost* true. Thus, these new theories suggest that protons have a half-life of 10^{31} years (see chapter 11), breaking down into nonbaryons such as positrons and neutrinos. It can do so even without the corresponding breakdown of antiprotons.

Similarly, it is possible for a baryon to be produced without the production of a corresponding antibaryon, though in a vanishingly small percentage of cases. As the temperature rises to dizzying heights, this asymmetric tendency grows more important, though it is never overwhelming.

Even at the many trillions of degrees of temperature under which the prequark particles are breaking down, the asymmetry would only be one part in a billion. For every 1,000,000,000 antiquarks formed, 1,000,000,001 quarks are formed.

This means that for every 1,000,000,000 antiquarks undergoing mutual annihilation with 1,000,000,000 quarks, one single and lonely quark is left over. The tiny percentage of leftover quarks is sufficient to produce enough nucleons to make up the stars and galaxies of our universe. It is a small, but crucial asymmetry (as I promised you at the beginning of the chapter).

At the present moment, there are indeed about a billion photons in the universe (formed from that primordial mutual annihilation) for every existing baryon. What's more, as nearly as we can tell, everything we know in the universe seems to be composed of baryons just about exclusively, with virtually no samples of antibaryons existing. (This was something that badly puzzled physicists when the law of conservation of baryon number was thought to be absolute.)

But why is it quarks, nucleons, and matter that are produced in slight excess, rather than antiquarks, antinucleons, and antimatter?

My own guess is that this asymmetry is inherent in the original holon and that there might also be an antiholon in which antiquarks would appear in slight excess so that an antiuniverse of antimatter would eventually be formed. (The photons of such an antiuniverse would be identical with the photons of a universe in all respects, however.)

If there are an infinite number of holons produced here and there in what I choose to call hyperspace, half should be holons and half antiholons. In fact, I suspect that they are produced in pairs, with every big bang actually a double big bang, producing a holon/antiholon pair. Thus, for our universe there is a corresponding antiuniverse which we cannot reach or impinge on in any way—nor it on us.

But where does the vast mass of the holon/antiholon pair come from? Does it suddenly appear out of nothing? Or are we forced, finally, to postulate a Divine Creator.

I prefer to wonder if the holon-antiholon pair might not actually appear out of nothing.

Suppose there is such a thing as "negative energy" which has the property of being able to combine with ordinary energy and cancel it out, leaving nothing. For every holon-antiholon pair there might be a negative-holon—negative-antiholon pair also formed; the two pairs together adding to nothing and therefore, mathematically at least, being equivalent to nothing. It would not be strange to have nothing appear out of nothing.

Thus, $(+1) + (-1) = 0$. If you have no money, you have nothing (fiscally speaking). If someone owes you a dollar and you owe someone else a dollar, you still have nothing—though you can now collect what is owed you and delay paying what you owe, so that you have a dollar to do something with.

In the same way, the quadruple holon is nothing, but the individual members can be played with for a time. The four holons appear out of nothing, each expanding in a quadruple big bang and playing out a temporary game as separate universes and each eventually contracting into a quadruple big crunch in which all the events of the expansion reverse themselves, right down to the final disappearance, into the nothingness from which they came. (And there may be an infinite number of such quadruple universes, coming and going.)

This, you will note, repeats the thesis of my essay "I'm Looking over a Four-Leaf Clover,"* which I still stand by. In the earlier essay, I had to rely on the dubious concept of antigravity to separate matter and antimatter. Now I can accept, instead, the crucial asymmetry inherent in GUT, which is a big improvement.

I must still make use of the dubious concept of antienergy, however. There I must wait longer for science to catch up.

15* ALL AND NOTHING

It was back in 1967, I think, that I read J. R. R. Tolkien's *The Lord of the Rings* for the first time. I liked it—moderately. I felt it went on too long; that there was too much irrelevant detail; that the battle scenes were a bit wearisome.

I have since read it four more times and have just finished the fifth reading.

Each time I liked it better than the time before and on this fifth occasion I clamored restlessly against having it end at all. Far from thinking it went on too long, I bitterly resented Tolkien's having waited so long to start it that he ended with only time to write half a million words.

It's fair to wonder why I should like it better each time I read it. After all, with each reading, the details of the plot are more firmly ground into my head and there is less chance of any suspense.

But then it's not the plot that counts. That can be summarized in a few pages and one is glad to have it over and done with. Once one gets to know the plot very well, one can ignore it and not be so concerned with following it that one misses the more subtle beauties. (Naturally this is only true of a book that is more than the sum of its plot.)

* In my collection *Science, Numbers and I* (Doubleday, 1968).

What pleased me more and more, each time I read it, was the intricate construction of the whole. In particular, I am pleased with the way in which the epic starts small, separates into two parts, then has those parts rejoin and end small.

What's more, of the two parts, one is a colossal war that grows more and more extensive and intensive until it encompasses the whole world and threatens all of it with eternal destruction, while the other has a focus that is ever-narrowed into a smaller and smaller compass until it ends with two small beings taking weary step after weary step up the side of a volcano.

From large to small we go, then from smaller to larger, then from still larger to still smaller—and in the end it is the small that counts. The apparent nothing saves the all.

Tolkien plays fair. He tells you all along that that's the way it will be, but telling you doesn't count. He *shows* you. And though I know that the nothing will save the all, and exactly how—each time I read the epic I appreciate and admire and enjoy the artistry of the technique more.

"All art," said Seneca, the Roman philosopher, "is but an imitation of nature."

Yes, but some aspects of nature are discovered after the construction of a particular piece of art, and if the art is valid, it will surely seem that nature imitates art.

For instance—

If we want to consider the "all" as far as the world of sense impressions is concerned, we can do no better than to consider the universe. There is the earth, which is part of the solar system, which is part of the galaxy made up of a hundred billion or so suns, which is in turn but one of a hundred billion or so galaxies.

And if we look at the universe, we find that one of its fundamental overall characteristics is that it is expanding. The galaxies exist in groups, some small and some large, and all of these groups are steadily moving away from each other. This expansion is, presumably, the result of an initial explosive event—the "big bang."

Given the rate of expansion it is possible to calculate the date of the beginning of the universe with some degree of confidence, as I pointed out in chapter 13.

But what about the other end of the arrow of time? What is the ending of the universe and how far off is it?

We might argue that though there is a limit to how far back in time we can go, since the universe gets smaller as we move back in time and must sometime reach zero volume, there is no limit to how far forward in time we can go. The further forward we go the larger the universe gets but it is difficult to see how there can be any maximum size past which it cannot go. It can expand forever and can, in that sense, have no ending.

To use a simple analogy, suppose the earth is alone in the universe and there is an object a mile above the surface of the earth that is moving upward at a constant speed. We can tell when it started its journey because if we consider the situation further and further back in time, we would be aware of the object's being located closer and closer to the surface of the earth. At some time in the past, the object would have been on the surface and that would represent the moment at which it began its upward journey.

However, if we consider the situation further and further forward in time, the object would be moving higher and higher and since there is nothing in the upward direction to stop it, we could conclude that it would travel forever and its journey would have no end.

An infinitely expanding universe is an "open universe." If we suppose that the universe is open we can see that eventually, after the passing of some vast time period, the various clusters of galaxies will be separated by such huge distances that no one cluster can be detected from any other. In that case, intelligent observers on earth (or some other equivalent vantage point) in the far future would be able to see only objects in our own galaxy or in the other galaxies of what is called the *local group*. Aside from that, the universe would seem empty.

There might be changes in the basic properties of the universe as a result of the continuing expansion. Some argue that the intensity of the gravitational force decreases as the universe expands. Less controversially, it is argued that the average temperature of the universe drops as it expands (it is only about three degrees above absolute zero right now—on the average).

If we ignore the possibility of such changes, we might conclude that the mere expansion of the universe in this fashion would lose ordinary people nothing, however much astronomers might regret the loss of the outer galaxies. After all, the only objects we can see in the sky at all, with the unaided eye, are the nearer stars of our own galaxy, plus such other members of the local group as the Magellanic clouds and the Andromeda galaxy.

Nor would we have need to feel hampered even if we possessed the capacity for easy interstellar flight. Our local group contains about a trillion stars and surely that is not an ungenerous figure. It would still leave us much to study.

But then, we can't expect our stars to remain unchanged. According to the second law of thermodynamics (see chapter 13), we know that available energy is going to come to an end someday. The stars will each consume its fuel and collapse. The smaller stars will collapse relatively quietly to white dwarfs, which will slowly cool off to black dwarfs.

Larger stars will explode and collapse to neutron stars or even to black holes. Eventually, white dwarfs and neutron stars would sweep up enough mass on their voyage through space to collapse further to black holes, while the black holes will become steadily larger and more massive.

It would seem then that all the mass of the universe might end up eventually as part of one black hole or another; that in place of each galaxy there might be an enormously massive black hole representing its core, possibly with planetary black holes of smaller size representing each a portion of its outskirts. These black holes would exist in clusters, large and small, representing the galactic clusters, and all the clusters of black holes would be forever receding from each other.

It is now believed, however, that black holes evaporate slowly. This evaporation proceeds more rapidly the smaller the black hole is, and the lower the average temperature of the universe is. Under present conditions the rate at which the typical black hole picks up matter from its surroundings far outweighs any evaporative tendency.

As we move forward in time, however, there will come a period when there is virtually no non-black-hole matter to absorb and when continued expansion will have lowered the average temperature of the universe many times closer to absolute zero than it now is. The evaporation of black holes will then dominate, and they will shrink slowly, producing matter that will spread out as a fine, thin cloud of dust, atoms, and subatomic particles, growing ever finer and thinner.

And that will be the end. Entropy will never be at quite a maximum, for it will continue to increase, though more and more slowly, as the universe continues to expand.

But I keep saying "eventually." How long is "eventually"?

The universe is very likely fifteen eons old (15 billion years, that is), although as I explained in chapter 13 there is some recent dispute about this. The small red dwarfs, however, which make up three-fourths of all the stars, and which use their hydrogen fuel stingily, can trickle out their lifetimes as normal stars over a period of two hundred eons. (Compare this with the sun's lifetime, as a normal star, of about twelve eons.)

This means that even those red dwarfs which were created in the infancy of the universe are still in their youth and have expended less than a tenth of their fuel store.

And new stars form continually, since there remains uncondensed gas and dust in the interstellar spaces and since still more is added to the supply continually as supernovas explode. To be sure, as old gas and dust condenses and as new supernova-born material is added, the hydrogen content of the gas and dust out of which new stars are made steadily decreases, while the heavy-element components steadily increase.

In the end, the fuel will be all gone, but I don't see how this can take less than a thousand eons.

And how long will it take for white dwarfs and neutron stars to grow cold? For all matter to find its way into one black hole or another? For all the black holes to evaporate so that the universe is nothing but a thin vapor growing ever thinner?

I don't know. I have never seen a reasonable estimate, but I suspect it would take an enormous number of years, quite beyond meaning. In such a universe the length of time during which life as we know it would be possible would be, comparatively, the merest instant of history—the instant, of course, in which we happen to be passing through.

I don't like the picture of an open universe. If that is how things are, I must accept it whether I like it or not—but is there possibly another interpretation of the universe?

Suppose we go back to the analogy of the object rising upward from the surface of the earth. I suggested that it was rising upward at a constant velocity, but that is impossible, of course. The rising object is subjected to earth's gravitational pull at all times and therefore its upward velocity is steadily being slowed. It is moving upward more and more slowly.

If earth's gravitational pull were constant with distance, then no matter how quickly the object was moving upward, its velocity would be bled away until it was zero. The object would, in other words, stop moving upward, halt momentarily, and then start falling.

Earth's gravitational pull weakens with distance, however. If the object is moving upward at more than a certain velocity, it increases its distance from earth's center so rapidly that the gravitational pull drops too quickly to slow that velocity effectively. The velocity, in that case, *never* decreases to zero with respect to earth and the object *never* falls back again. The minimum velocity at which this happens is the "escape velocity."

At velocities lower than escape velocity, the object moving upward will eventually slow to a momentary halt and begin to fall (though it will attain a greater height than it would if earth's gravitational pull did not decrease with distance).

The expanding universe is much in the position of the object moving upward from earth's surface. The universe *cannot* expand at a uniform rate with time because it is expanding against the pull of its own overall gravitational field. The rate of expansion must, therefore, be slowing.

The question, then, is whether the rate of expansion is above or below the escape velocity of matter with respect to the overall gravitational field of the universe? If it is above the escape velocity then the universe is open and will expand forever. If it is below the escape velocity then we have a "closed universe," one that will expand at a slower and slower rate until it comes to a momentary halt and will then start contracting again.

A contracting universe will occupy a smaller and smaller volume with time, until a volume is reached which is so small that it will trigger a new big bang. The process will then continue forever into the future and will have been continuing forever in the past—like an object falling and bouncing, rising to a slowing halt, then falling and bouncing, rising to a slowing halt, and so on.

Nor is anything lost between bounces, or big bangs. Not only is all the matter of the universe retrieved in the course of the contraction, but all the immaterial radiation as well.

Radiation does not move in a straight line. If it did we might imagine it moving forever outward and escaping regardless of whether the matter of the universe were expanding or contracting. Instead, radiation follows the curve of an Einsteinian universe, with the presence of matter producing the curves to varying degrees of tightness. In an open universe, such a curved path, whatever its local veering, spirals outward indefinitely on the whole. In a closed universe, on the other hand, such a curved path spirals outward to a definite limit, then spirals inward again.

In a closed universe, there is no true beginning, no true ending. The universe repeats itself endlessly and we can only speak of the beginning and ending of a particular oscillation (the "wavelength of the universe," so to speak). How long an interval there is between big bangs depends on how intensely closed the universe is, and my own guess is that if the universe were closed, the interval might approach a thousand eons. Just a guess.

I feel attracted to the notion of the closed universe. It is cyclic and neat and it offers a new chance for life at each oscillation. Therefore, I want it to exist.

Unfortunately, a closed universe won't exist just because I want it to. Whether it exists or not depends on what the evidence tells us; in other words on whether the present rate of expansion is above the universal escape velocity or below it.

Given a particular rate of expansion, the escape velocity depends on the intensity of the general gravitational field of the universe, which depends, in turn, on the average density of matter in the universe.

Thus, *if* the rate of expansion is such that a galaxy's speed of recession increases by 50 kilometers per second for every million parsecs additional distance from us (a generally accepted figure despite recent questioning of it), then the average density of the universe must be about 5×10^{-30} grams per cubic centimeter as a minimum for the universe to be closed.

The volume of the universe, at present, is about 10^{85} cubic centimeters (a 1 followed by 85 zeroes). If the universe had just the minimum average density required for it to be closed, its total mass would then be $5 \times 10^{-30} \times 10^{85}$, or 5×10^{55} grams, or 2.5×10^{22} times the mass of our sun. For a closed universe, there would have to be the equivalent of 25,000,000,000,000,000,000,000 (25,000 million trillion) sun-sized stars in the universe.

The trouble is that the average density of the universe, judging from the mass of the average galaxy and the average distance between galaxies, is no more than one one-hundredth that minimum, so that there are the equivalent of only 250 million trillion sun-sized stars in the universe.

That means that the gravitational field of the universe is only one one-hundredth as intense as it ought to be to be able to bring the universal expansion to an eventual halt. Therefore, the universe is open and this period of life potentiality is and will be (as far as we know) the only one ever.

Can we possibly be wrong? Can the average density of the universe be higher than we think it is—a hundred times higher at least?

It might occur to you, perhaps, that the universe may be a hundred times larger than we think it is; that if we had better telescopes we could see five times farther out and see more and more galaxies.

That wouldn't work. The trouble would be that the new mass would be occupying new volume, and the overall density of the universe (which is mass divided by volume) would not change.

What we need is additional mass *within the volume we now observe*. This is the problem of the "missing mass."

You might think that there is no problem. The missing mass just isn't there and the universe is open, and just because Asimov wants the universe to be closed isn't going to produce the mass.

Well, I have faith in the beauty of nature and whereas a closed universe is beautiful, an open universe is not (and I have committed myself to a closed universe in print over a number of years). Besides there is the matter of galactic clusters.

If the mass and volume of a galactic cluster is determined (as nearly as possible) and if the average density of matter over the region occupied by that cluster is calculated, it almost always turns out that the general gravitational field of the cluster is not sufficient to keep the individual galaxies within the cluster. What is more, the larger the cluster, the greater the percentage by which the gravitational field falls short.

And yet the galactic clusters seem to be definitely closed. As nearly as we can tell they are not breaking up but are holding together. The missing mass, in galactic clusters at least, cannot therefore really be missing, if there is any meaning to celestial mechanics. The missing mass must be there somehow even though we don't see it, and if it is true of the galactic clusters, it must be true of the universe as a whole.

But where can the mass be? Of course, we can't always see massive objects. Stars shine, but black holes don't, and a black hole without any matter nearby to fall in and release X rays might have a wholly unexpected existence, and we might all be innocently overlooking its mass.

Suppose, for instance, there were very massive black holes at the center of each galaxy; black holes so massive that each would be far more massive than the galaxy itself. In that case, might not each galaxy be a hundred times more massive than we expect, the density of the universe a hundred times greater than we think, and the universe closed?

But that is not the answer. We don't estimate the mass of a galaxy by the number of stars we see in it so that it is not a case of neglecting the invisible. The mass of a galaxy can be determined, for instance, from its diameter and its speed of rotation. The rate at which its outer stars wheel depends very largely on the mass of the galactic core (which makes up about 90 percent of the mass of the whole). In this way we determine the mass reliably, black holes and all. The hidden mass cannot be lurking in black holes at the galactic core.

In that case, suppose we look outside the cores. Suppose that the galaxies themselves are larger than we imagine; that they stretch out in a thin powdering of stars and dust and gas—a kind of "galactic atmosphere." The usual methods for determining galactic masses ignore the possibility of such galactic atmospheres, but although they are thinly spread out they might fill the vast spaces between the galaxies and multiply the total mass, and density, a hundredfold.

There is, in fact, some indication of a thin powdering of stars, dust, and gas stretching out beyond the visible rims of galaxies. This does supply additional mass but, as far as we can tell, not nearly enough. They might add one-tenth the known mass, but surely not one hundred times the known mass. If the galactic atmospheres were dense enough to do that, they would be sufficiently noticeable to leave us in no doubt.

But again, let us go back to black holes—not at the core this time, but spread out through space generally. If they and their mass went generally unnoticed and could not be detected by studies of galaxies themselves, might that be the answer?

Possibly, but since we don't see such black holes or detect them in any way, or have any evidence whatever that they're there, it is difficult to feel confidence in this as a solution. In fact, there is indirect evidence on the basis, for instance, of the quantity of deuterium (hydrogen-2) in the universe, that the overall mass of the universe and, hence, the overall density, can't be much more than we think.

Well, then, having carried the "all" as far as we can go, let's turn to the "nothing" portion of the argument. The nearest we can get to nothing is, of course, the subatomic particles.

The mass of the universe is made up very largely of protons, neutrons, and electrons. If there is antimatter, that is made up of antiprotons, antineutrons, and antielectrons. There could be a wide variety of other particles, both leptons and hadrons, which are formed as a result of energetic events here and there in the universe.

The energetic particles are so few in number, however, that they don't contribute significantly to the total mass. The electrons are many in number but so light in comparison to protons and neutrons (the mass of 1 proton is equal to that of 1,836 electrons and 1 neutron to that of 1,838 electrons) that they don't contribute significantly. We have no evidence that antimatter exists in significant quantities in the universe, so we can eliminate antiprotons and antineutrons.

In short, we can say that the mass of the universe is, essentially, the total mass of the protons and neutrons it contains, and we can speak of those protons and neutrons together as nucleons.

The mass of a single nucleon is equal to 1.66×10^{-24} grams. If the mass of the universe is 5×10^{53} grams (one one-hundredth the amount required for closing), then the total number of nucleons in the universe is 3×10^{77}. We need a hundred times that many.

Well, then, are there any particles with mass that we've neglected? We've dismissed the electrons, the energetic particles, the antiparticles. What else?

Actually, there are particles present in the universe in numbers greater than the nucleons; far greater. The catch is that these particles are thought to have zero rest mass and they do not therefore contribute to the general mass of the universe, to its overall gravitational field, and to its closing.

These zero-mass particles come in three varieties: gravitons, photons, and neutrinos.

Gravitons, the exchange particles of the gravitational interaction, have not yet been detected. Physicists are certain they exist, however, and will be detected as soon as we can make our detecting devices delicate enough.

Photons, the exchange particles of the electromagnetic interaction, are very easy to detect. We do it with our eyes, if nothing else.

Being exchange particles, gravitons and photons both interact readily with matter and are constantly being absorbed and reemitted. On the other hand, neutrinos, the third variety of zero-mass particle, are *not* exchange particles and scarcely interact with matter at all. They pass through us, all of earth, all of the sun, without a swerve, without any indication of being aware we or it exist. One neutrino out of many trillions may hit an atomic nucleus sufficiently head-on to be absorbed, but no more than that. (However, that's enough to keep them from going utterly unnoticed.)

It is the neutrino, without mass, without electric charge, without interaction with other particles, that is the nearest thing to a nothing particle we know.

It is to this "nothing" then that we turn for a solution to the mystery of the "all," of the missing mass whose presence or absence will make the difference between an open and a closed universe, an eternally expanding one, or an oscillating one, one in which life exists for virtually zero time or one in which life can arise and arise, again and again, eternally.

And we'll go on with the subject in the next chapter.

16* NOTHING AND ALL

Back in 1955, when I was a full-time and active member of the faculty at the medical school, I was promoted to the rank of associate professor of biochemistry, something which swelled my bosom with pride, you may be sure.

Over the course of the next three years, however, I was engaged in a Homeric struggle with the director of the institution* and, as of June 30, 1958, I was no longer a full-time, active member of the department. I did manage to retain the title, just the same, and I made it perfectly clear to everyone that I didn't intend *ever* to give it up.

So there I was, without duties and without salary, but still associate professor of biochemistry.

The years passed, the decades passed, and my title remained—unchanged. And I grew sad for I was becoming a little too intensively late in my youth for that qualifying adjective, and I knew there was no way of getting promoted as long as I wasn't really teaching.

But things were changing at the university, too, and the attitude

* The details are in *In Joy Still Felt* (Doubleday, 1980), the second volume of my autobiography.

toward me altered continually for the better. Finally, I was told they were going to promote me.

I said cautiously, "But I'm not doing any teaching and I am no longer in a position to do more than give an occasional lecture."

They explained the situation to me in words of one syllable. "Who cares?" they said.

And as of October 1979, after twenty-four years, I became (finally) a professor of biochemistry. No adjective.

By a peculiar coincidence, the same thing happened to the neutrino. Discovered just after I became associate professor, it, too, had to wait twenty-four years before being promoted to a new level of importance.

The difference is that my promotion is definite, while that of the neutrino is still very tentative. On the other hand, if the neutrino's promotion holds, the results are of truly cosmic importance, whereas my own promotion is perhaps a hair's breadth short of cosmic.

But let's get to the details.

The existence of the neutrino was first predicted on theoretical grounds in 1931 by the Austrian-American physicist Wolfgang Pauli. Its properties, if it was to meet the theory, made it elusive. It had no electric charge and little or no mass, and did not interact with matter. Under those circumstances, it was a "nothing particle" and the chances of detecting it were almost nil.

It was not until 1956 that two American physicists, Frederick Reines and Clyde L. Cowan, set up an experimental procedure that definitely detected the neutrino and demonstrated its existence. (Actually, there are both a neutrino and an antineutrino, the two being equal and opposite in certain properties, and it was the antineutrino that Reines and Cowan detected. For purposes of this chapter, however, we'll let *neutrino* stand for both.)

Although neutrinos slip through matter without any trouble, so that many trillions pass through the earth from end to end for every one that is stopped on the way, they are not to be ignored, if only because they exist in such quantity.

The sun produces them copiously, for instance. The sun owes its existence as a light-radiating body to the energy derived from the conversion of hydrogen atoms to helium atoms. Four hydrogen atoms are converted to one helium atom, and for every helium atom produced, two neutrinos are also produced. Since hundreds of millions of tons of hydrogen are converted to helium every second in the sun, you can well see that enormous quantities of neutrinos are boiling out of it constantly.

In fact, 1.75×10^{38} neutrinos emerge from the sun every second. That's 175,000,000,000,000,000,000,000,000,000,000,000,000, or 175 trillion trillion trillion neutrinos. Since only a negligible number of neutrinos interact with other subatomic particles and in this way lose their identity, we are not too far off if we say that all the neutrinos produced by the sun are permanent residents of the universe.

The sun has existed for something like 1.5×10^{17} seconds, and if it has been producing neutrinos at its present rate for almost all that time, the universe contains 2.6×10^{55} neutrinos that are sun-born.

Then, too, there are several hundred billion stars in our galaxy and perhaps a hundred billion galaxies altogether, and the universe as a whole may be three times as old as the sun, so that altogether the number of neutrinos is far more enormous still.

If all the stars in all the galaxies had been producing neutrinos at the same rate as our sun does (on the average) throughout the lifetime of the universe, the total number of neutrinos would be something like 10^{78}. This, however, would be an underestimate, for it is quite likely that the major neutrino producers are the relatively few stars of great mass, the supernova explosions and other violent events, and perhaps even the big bang itself.

At any rate, astronomers estimate that if all the neutrinos now existing in the universe were to be spread out evenly, there would be 100 in every cubic centimeter. (Of course, they would only be passing through, for neutrinos, which are thought to be massless, would be traveling at the speed of light.)

Since more neutrinos are constantly being formed and virtually none are destroyed, the number per cubic centimeter is constantly increasing, but slowly. If that number is now exactly 100, and if the universe is 15 billion years old and has been producing neutrinos at the same rate all that time, then it will not be for another 150 million years that the density will rise to 101 neutrinos per cubic centimeter.

The universe, assuming a radius of about 12.5 billion light-years, has a volume of roughly 10^{85} cubic centimeters, which means it contains about 10^{87} neutrinos. For every neutrino in the universe that has been manufactured by our sun, there are a hundred million trillion trillion neutrinos manufactured by other bodies.

In the previous chapter, I said that the total number of nucleons (protons or neutrons) in the universe is 3×10^{77} and that these were thought to account for some 99.9 percent of the mass of the universe. It follows that there are roughly 3 billion neutrinos in the universe for every nucleon, but if neutrinos have zero rest-mass they do not, naturally, contribute to the total mass of the universe, except for their energy content which is excessively small.

There is another class of common zero-rest-mass particle—the photon. This is the constituent particle of electromagnetic radiation (of which visible light is the best-known example). Like the neutrinos, photons are massless (except for energy equivalence) and chargeless, but unlike the neutrinos they interact readily with matter so that they are forever being absorbed and reradiated. It is estimated that the universe contains a billion times as many photons as it does nucleons.

Well, then, if all the matter in the universe were smeared out perfectly evenly and if we took a snapshot of the resulting mess in a zero-time instant, it would turn out that in every volume of 30 cubic meters, there would be 3 billion neutrinos, 1 billion photons, and 1 (mark you, *one*) nucleon.

And yet since it is only that 1 nucleon that is thought to have mass, it is only that 1 nucleon that plays a role in deciding whether the universe is open or closed, whether it will expand forever, or whether it will someday bring its expansion to an ever-slowing halt and begin to contract again. Most astronomers feel that the 1 nucleon is not enough to close the universe and to make it contract someday. There should be more like 100.

The most common method of neutrino production is that of the conversion of a proton to a neutron. This takes place by the decillions of decillions per second the universe over, in the course of fusing hydrogen into helium in the various stars. As a proton changes to a neutron, a positron (which is an *antielectron*) and a neutrino are formed.

In the much less frequent, but equally possible, change of a neutron to a proton, an electron and an antineutrino are formed. We have agreed, for convenience' sake, to let the term *neutrino* include both itself and its antiparticle. Let us do the same for *electron,* which will include both itself and its antiparticle, the positron.

If we do this, we can say that neutrino formation and electron formation usually go together. In a way, this is fitting because both electrons and neutrinos belong to a class of particles called *leptons,* and the formation of both types of leptons according to particular rules makes it possible to have a "conservation of lepton number," to say nothing of preserving several other conservation laws. (It was in order to preserve these laws that Pauli suggested the existence of the neutrino in the first place.)

But then, other leptons were discovered. In 1935, nearly forty years after the discovery of the electron, the American physicist Carl D. Anderson discovered the *muon.* The muon is 207 times as massive as the electron, but in all other properties it is identical with the electron. Nowadays, we speak of the muon as another "flavor" of the electron.

We might picture the situation thus: An electron is in an energy valley. If energy is pumped into it, it is driven up a mountain slope until it finds a ledge in which it can rest momentarily. It is then a muon. However, the rest is only momentary. The muon is unstable and in a millionth of a second or so breaks down to an electron again, giving off energy in the form of neutrinos.

There are, of course, a muon and an antimuon—analogous to the case of the electron. The formation or decay of a muon involves the production of an antineutrino, and the formation or decay of an antimuon involves the production of a neutrino—again analogous to the case of the electron (except that the electron is stable and doesn't decay).

The neutrinos associated with muons seem to be identical in all properties with those associated with electrons and yet it seems they do not substitute for each other. If we speak of electron neutrinos and muon neutrinos, one can participate in certain specific subatomic particle-interactions that the other cannot.

In fact, physicists speak of "conservation of electron number" and "conservation of muon number" as two separate laws and if both are to exist, the electron neutrino and muon neutrino must be distinct and must exist as separate flavors.

Recently, a third flavor of the electron was detected—a still higher and more energetic ledge on the mountainside. This is the *tau electron* (which should be called the tauon, to my way of thinking). It has associated with it a tau neutrino, which is, again, indistinguishable from the electron neutrino and the muon neutrino, but is somehow a separate particle. (There may be an infinite number of ledges, higher and higher up the mountainside, requiring greater and greater energies to form. Presumably each progressively higher ledge is less likely to be occupied and plays a less important role in the universe. In any case, we are dealing with only three flavors of the electron so far, and three flavors of the neutrino.)

The existence of three different flavors of neutrino, indistinguishable but distinct, is troublesome. If they are distinct, they must differ in some property we have not yet learned to measure or even perhaps to recognize. And perhaps the difference is so subtle that it is not a true difference.

For instance: All particles have properties that can be associated with wave forms (just as all waves have properties that can be associated with particle forms). Therefore we can suppose that neutrinos travel through the universe in the form of tiny waves. What if there are three kinds of waves—one for the electron neutrino, one for the muon neutrino, and one for the tau neutrino—an e-wave, an m-wave, and a t-wave?

Suppose, further, that each neutrino is made up of all three types of waves, with one of them dominant. The electron neutrino, for instance, has the e-wave dominant, with the m-wave and the t-wave supplying minor modifications. The wave summation is a slightly altered e-wave, but one that, despite the alteration, is still recognizably an e-wave. We would then have an electron neutrino.

The same argument, suitably modified, can be used to define a muon neutrino and a tau neutrino.

If all three waves traveled at the same speed, the wave summation would always be the same and a particular flavor of neutrino would remain that flavor forever.

But what if the three waves traveled each at a different speed?

In that case, the relation of each wave to the others would constantly change, as one wave constantly moved ahead of the second, and as the third constantly fell behind the second. The summation wave would change in some regular fashion, with first one variety of wave and then another dominating. In that case, a neutrino would be sometimes an electron neutrino, sometimes a muon neutrino, sometimes a tau neutrino. This situation is referred to as *neutrino oscillations*.

This was first suggested as a possibility in 1963 by a group of Japanese physicists.

In the late 1970s, Frederick Reines, one of the original detectors of the neutrino, along with Henry W. Sobel and Elaine Pasierb of the University of California set up an experiment designed to see if such oscillations took place. For the purpose, they used 268 kilograms of very pure heavy water, containing hydrogen nuclei made up of a proton and a neutron in close association. They shielded the heavy water so that no particles could get in but neutrinos and used a uranium-fission reactor as a neutrino source. The reactor produced *only* electron neutrinos.

A few of the neutrinos, of the many trillions formed, strike the heavy hydrogen nuclei and two interactions are possible. First, the neutrino, on striking the proton-neutron combination, will simply split it apart and will then keep on going itself. This is a "neutral-current" reaction and *any* of the neutrino flavors can bring this about. Second, the neutrino, on striking the proton-neutron combination, can induce a change of the proton into a neutron, producing an electron, and in this case the neutrino ceases to exist. This is a "charged-current reaction" and *only* an electron neutrino can do this.

If electron neutrinos remain electron neutrinos at all times, physicists can calculate exactly how much of each reaction ought to take place in a given time. If there is oscillation, so that the electron neutrino is sometimes a neutrino of other flavors, then the neutral-current reaction isn't affected, but the charged-current reaction should take place considerably less often than is to be expected, since when the electron neutrino was in other flavors it could not produce a charged-current reaction.

Reines found that the neutral-current reaction was quite up to the proper level, but that the charged-current reaction took place with somewhat less than half the calculated frequency. He therefore announced, in 1980, that his experiment seemed to demonstrate the existence of neutrino oscillation.

I say "seemed" because the experiment was carried out at the limit of the detectable. Only about 80 neutrino events per day were detected that belonged to the experiment, among 400 neutrino events that seemed to be the result of neutrino bombardment from sources other than the fission reactor.

Besides, other experiments, conducted in Geneva, Switzerland, did not seem to show neutrino oscillation. What will be needed, therefore, will be more and better experiments, and I have no doubt they will be forthcoming. There is even a chance that events will outdate this book before it sees print.

What makes the possibility of oscillation persuasive to me, despite the tenuousness of the evidence, is that it explains several different, apparently unrelated, puzzles in astronomy.

First, there is the matter of the solar neutrinos. For years now, astronomers have been puzzled over the fact that careful attempts at detection have not uncovered as many neutrinos issuing from the sun as theory would lead them to expect. In fact, the highest figure obtained is about a third the theoretical.

The neutrino-detecting devices, however, are designed to detect *only* electron neutrinos, and the supposition was that the sun emitted only electron neutrinos and that these remained electron neutrinos all the way to the earth.

If, however, there are oscillations, what reaches us could be an equal mixture of the three flavors and then we would detect only one-third the electron neutrinos we would expect to detect.

Here's another point. If all three flavors of neutrino possessed zero rest-mass (as the assumption has been for decades), then all three must travel through the vacuum of space at the speed of light. Traveling at equal speeds, the neutrinos could not oscillate, for that depends on the three waves traveling at different speeds. If, however, the neutrinos had a very tiny mass, they would travel at a trifle less than the speed of light, and if each flavor had a slightly different tiny mass (that representing the hitherto unrecognized distinction among them), they would each travel at a slightly different speed just short of that of light.

In other words, neutrino oscillation implies that at least one of the neutrino flavors has mass, and perhaps all three. There is, in fact, a report of experiments by physicists in Moscow that involves a point that has nothing to do with oscillations and that seems to show the electron neutrino to have a mass of possibly as much as 40 electron volts. This would give it a mass $1/_{13,000}$ that of an electron or about $1/_{23,000,000}$ that of either a proton or a neutron. This tiny amount is much greater than the mass equivalent of a neutrino's energy.

Even so, that's not much of a mass, but then there are a heck of a lot of neutrinos.

For instance, I said there would be 100 neutrinos per cubic centimeter if all the neutrinos were spread out evenly over the universe. However, they are *not* spread out evenly.

It is in the core of the stars that the vast majority of neutrinos are formed, and even though neutrinos move at very nearly the speed of light, it would take considerable time for them to get really far away from the stars in which they were formed. It would take even more time for them to get really far away from the galaxy in which they were formed, and even more time to get away from the galactic cluster in which they were formed.

In other words, one would expect a cloud of neutrinos to exist in the neighborhood of a galaxy and, even more so, in the neighborhood of a cluster of galaxies. The cloud would be a permanent one, for even as the neutrinos disperse in all directions at nearly the speed of light, so that the cloud forever thins out at its edges, new swarms of neutrinos are constantly being formed as replacements by all the stars of the galaxies.

This sounds pretty good, but there is a catch. If the neutrinos are produced by the stars, their mass is subtracted from the stars. That means if a large quantity of the mass of the galaxies is in the neutrinos the stars have produced, then originally, before there was time for many of the neutrinos to be formed, the stars themselves must each of them have been far more massive than they are now, and that is not likely.

The stars could not have begun as much more massive than they are now or there would have been serious consequences in the course of stellar evolution, consequences that have not been observed. Therefore, the star-produced neutrinos cannot actually add much in the way of mass to galaxies and clusters of galaxies.

It may be, though, that by far the great majority of the neutrinos now in existence were produced at the time of the big bang. They are primordial neutrinos. If they had zero rest-mass, they would spread themselves out through the universe more or less evenly. If they had mass, even a slight mass, they would be trapped by the gravitational field of galaxies and collect there in vast numbers without our suspecting they were there. And *then* they would account for the missing mass.

Let's consider that more slowly.

Astronomers associate the mass of stars with their luminosities in a fairly straightforward manner. If the mass of an entire galaxy is considered, however, the overall luminosity of the galaxy is less than might be expected from the relationship of mass and luminosity existing among individual stars. Clusters of galaxies fall short in luminosity to an even greater extent.

It is almost as though a substantial fraction of the mass of a galaxy were deficient in luminosity, or were perhaps entirely dark; and that this was even more extreme in clusters of galaxies. The existence of such "dark mass" is also indicated by the fact that clusters of galaxies, where the dark mass is greatest, do not seem to have the required intensity of gravitational attraction, judging by its stars alone, to hold the individual galaxies of the cluster in place. The dark mass is necessary, then, for the very existence of the clusters, as clusters.

One possibility is that the dark mass consists of black holes, but there is no good evidence for the existence of enough black holes to account for it. This is especially so since black holes in the quantity required ought to give themselves away by the effects of their gravitational fields, which are enormously intense in their near vicinity.

But now a second possibility arises. What about the neutrino cloud that hovers about galaxies and clusters of galaxies? Even if each neutrino had an almost negligible mass, the total would be enough to account for the shortfall in luminosity and for the manner in which clusters of galaxies remain together.

And that brings up a third problem, the most cosmic of all: the question as to whether the universe is open or closed.

I have already said that for every nucleon in the universe there seem to be about 1 billion photons and 3 billion neutrinos.

There seems to be no question (at least so far) that photons are truly without rest-mass. Therefore the mass of the photons is only that equivalent to their energy. That is very small and they do not contribute significantly to the possible closing of the universe.

On the other hand, suppose that neutrinos do oscillate and that they therefore have rest-mass and suppose that the Soviet estimate of the rest-mass of the neutrino as equal to 40 electron volts is correct. In that case, the 3 billion neutrinos that exist for every nucleon have a total mass equal to 125 nucleons.

The startling conclusion is that the mass of the universe is not primarily that of the nucleons at all, as has been supposed. If neutrinos do, in fact, oscillate, then $^{125}/_{126}$, or 99.2 percent of the mass of the universe, is due to its neutrino content, and it is these "nothing particles" that are essentially the universe. Everything else—all the matter that makes up the black holes, quasars, pulsars, stars, planets, cosmic dust, and us—is just an inconsiderable impurity.

What's more, astronomers have said that in order for the universe to be closed, there would have to be about one hundred times as much mass as careful observations seemed to indicate it has. That careful observation, however, has been devoted entirely to determining, in effect, the quantity of nucleons in the universe.

But, if the neutrinos oscillate then it is quite likely that the total mass of the universe is more than a hundred times the mass of its nucleons and by a comfortable margin, too. That means the universe is closed; that its expansion will someday stop and that it will then begin to contract; that there will eventually be a new big bang; and so on, over and over, world without end.

And I must admit that, emotionally, that's what I want. I think a closed universe is elegant and beautiful compared with a one-shot open universe that expands without end and dies without progeny. I earnestly hope, therefore, that Reines's results prove to be correct, and while, if they do not, I will have to accept that, I will do so without joy.

I would like to point out, too, that if neutrinos close the universe, then we have a replay on an infinitely more cosmic scale of the drama of *The Lord of the Rings* (to which I referred at the beginning of the previous chapter).

The cosmic fate of the all (the universe) is decided at last by the tiny mass of the nothing (the neutrino); just as the worldwide struggle against the evil of Sauron was decided at last by the painful delivery of the Ring to the Mountain of Doom by little Frodo.

Nature imitates art.

D ★ LITERATURE

17★ MILTON! THOU SHOULDST BE LIVING AT THIS HOUR

Some time ago I was signing books at Bloomingdale's department store. I don't recommend this as a general practice if you are in the least bit shy or sensitive.

It involves sitting at some makeshift table with a pile of your books about you, amid a vast display of women's garments (that happened to be the section near which I was placed). People pass you with expressions varying from complete indifference to mild distaste. Sometimes they look at the books with an expression that might be interpreted as "What junk is this that meets my eye?" and then pass on.

And, of course, every once in a while someone comes up and buys a book, and you sign it out of sheer gratitude.

Fortunately, I am utterly without self-consciousness and can meet any eye without blushing, but I imagine that those who are more sensitive than I would experience torture. Even I would give it a miss were it not that my publisher arranges such things and I don't want to seem unreasonably uncooperative in measures designed to sell my books.

At any rate, there I was at Bloomingdale's and a tall woman, in her thirties (I should judge) and quite attractive, rushed up smiling, with a pretty flush mantling her cheeks and said, "I am *so* glad and honored to meet you."

"Well," said I, becoming incredibly suave at once, as I always am in the presence of attractive women, "that is as nothing to my pleasure in meeting *you.*"

"Thank you," she said, then added, "I want you to know I have just seen *Teibele and the Demon.*"

That seemed irrelevant, but I said the polite thing. "I hope you enjoyed it."

"Oh, I *did.* I thought it was wonderful, and I wanted to tell you that."

There was no real reason for her to do so, but politeness above all. "That's kind of you," I said.

"And I hope you make a billion dollars out of it," she said.

"That would be nice," I admitted, though privately I didn't think the owners of the play would let me share in the proceeds even to the extent of a single penny.

We shook hands and separated, and I never bothered to tell her that I was Isaac Asimov and not Isaac Bashevis Singer. It would merely have embarrassed her and spoiled her kindly good wishes.

My only concern is that someday she will meet Isaac Bashevis Singer and will say to him, "You imposter! I met the *real* Isaac Bashevis Singer and he's young and handsome."

On the other hand, she may not say that.

But then, it's easy to make mistakes.

For instance, most people who have heard of John Milton think of him as an epic poet second only to Shakespeare in renown and genius. As evidence, they point to *Paradise Lost.*

I, on the other hand, always think of Milton as something more than merely that.

The poet William Wordsworth, back in 1802, found himself in low spirits when he decided that England was a fen of stagnant waters and moaned, "Milton! thou shouldst be living at this hour."

Well, Bill, if Milton were living at *this* hour, here in the late twentieth century, I'm sure he would be that acme of art, a science fiction writer. As evidence, I point to *Paradise Lost.*

Paradise Lost opens as Satan and his band of rebellious angels are recovering in hell, after having been defeated in heaven. For nine days the stricken rebels have been unconscious, but now Satan slowly becomes aware of where he is (and, if you don't mind, I will quote without the use of lines of poetry, but as simple prose, to save space):

"At once, as far as Angel's ken, he views the dismal situation waste and wild, a dungeon horrible, on all sides round as one great furnace flamed; yet from those flames no light, but rather darkness visible served only to discover sights of woe."

Milton is, essentially, describing an extraterrestrial world. (As Carl Sagan has remarked, our present view of the planet Venus is not very far removed from the common conception of hell.)

The remark about "darkness visible" is surely modeled on the description of Sheol (the Old Testament version of hell) in the Book of Job: "A land of darkness, as darkness itself; and of the shadow of death, without any order, and where the light is as darkness."

Milton's phrase makes it graphic, however, and is a daring concept, one that is a century and a half in advance of science; for what Milton is saying is that there can be some radiation that is not visible as ordinary light and yet can be used to detect objects.

Paradise Lost was published in 1667, but it was not until 1800 that the German-English astronomer William Herschel (1738–1822) showed that the visible spectrum did not include all the radiation there was; that beyond the red there was "infrared" radiation that could not be seen but could be detected in other ways.

In other words, with remarkable prescience, Milton had hell lit by flames that gave off infrared light, but not visible light (at least we can interpret the passage so). To human eyes hell would be in darkness, but Satan's more-than-human retina could detect the infrared and to him it was "darkness visible."

Where is the hell that is occupied by Satan and his fallen angels? The common view of the location of hell, from ancient times on, is that it is somewhere deep in the earth. The fact that bodies are buried underground contributes, I suppose, to this feeling. The fact that there are earthquakes and volcanoes gives rise to the thought that there is activity down there and that it is a place of fire and brimstone. Dante placed hell at the center of the earth, and so would most unsophisticates of our culture today, I think.

Milton avoids that. Here's how he describes the location of hell:

"Such place Eternal Justice had prepared for those rebellious; here their prison ordained in utter darkness, and their portion set, as far removed from God and light of Heaven as from the centre thrice to the utmost pole."

It is logical to suppose the "centre" to be the center of the earth, since that was also taken to be the center of the observable universe in the Greek geocentric view of the universe. This view was not shaken until Copernicus' heliocentric theory was published in 1543, but the Copernican view was not instantly accepted. Scientific and literary conservatives held to the Greek view. It took Galileo and his telescopic observations in 1609 and thereafter to establish the sun at the center.

Milton, however, although writing a good half-century after the discoveries of Galileo, could not let go of the Greek view. After all, he was dealing with the biblical story, and the biblical picture of the universe is a geocentric one.

Nor was this because Milton did not know about the telescopic findings. Milton had even visited Galileo in Italy in 1639 and refers to him in *Paradise Lost.* At one point, he finds it necessary to describe Satan's round and gleaming shield. (All the characters in *Paradise Lost* act and talk as much like Homeric heroes as possible and are armed just as Achilles would be; that's part of the epic convention.)

Milton says that Satan's shield is like the moon, "whose Orb through optic glass the Tuscan artist views . . . to descry new lands, rivers or mountains in her spotty globe." There is no question but that the "Tuscan artist" is Galileo.

Nevertheless, Milton doesn't want to be involved in astronomical controversy and in book VIII of the epic he has the archangel Raphael respond to Adam's questions on the workings of the universe in this way:

"To ask or search I blame thee not, for Heaven is as the Book of God before thee set, wherein to read his wondrous works, and learn his seasons, hours, or days, or months, or years: This to attain, whether Heaven move or Earth, imports not, if thou reckon right; the rest from man or angel the great Architect did wisely to conceal, and not divulge his secrets to be scanned by them who ought rather admire."

In other words, all that human beings need out of astronomy is a guide whereby to form a calendar; and to do this, it doesn't matter whether earth moves or the sun. I can't help but feel this to be a very cowardly evasion. Some of the pious of the world were willing to denounce, excommunicate, and even burn those who claimed the earth moved—until the evidence began to show clearly that earth *did* move, and when that came about, they then said, "Oh, well, it imports not; what's the difference?" If it "imports not," why did they make all that fuss earlier?

So the Miltonic universe remains geocentric—the last important geocentric universe in western culture. The "center" Milton speaks of in locating hell is the center of the earth.

The distance from the center of the earth to its pole, either the North Pole or the South Pole, is 4,000 miles, and this figure was known to Milton. The earth had been several times circumnavigated by Milton's time and its size was well known.

In that case, "thrice" that distance would be 12,000 miles, and if this is the interpretation of "from the centre thrice to the utmost pole," then hell would be 12,000 miles from heaven.

It seems reasonable to suppose that earth is equidistant from hell and heaven. If, then, heaven were 2,000 miles from earth in one direction and hell 2,000 miles from earth in the opposite direction, that would allow hell to be 12,000 miles from heaven if we count in earth's 8,000 miles diameter.

But this is ridiculous. If heaven and hell were each 2,000 miles away, surely we would see them. The moon is 240,000 miles away (something the Greeks, and therefore Milton, knew) and we see it without trouble. To be sure, the moon is a large body, but surely heaven and hell would be large also.

Something is wrong. Let's reconsider:

The line in Milton reads, "from the centre thrice to the utmost pole." What is the *utmost* pole? Surely, it is the celestial pole, the point in the sky that is directly overhead when you stand on a terrestrial pole.

No one in Milton's time knew how far away the celestial pole was. Astronomers knew the moon was a quarter of a million miles away and the best guess the Greeks had been able to make where the distance of the sun was concerned was 5 million miles. Since the sun was the middle planet of the seven (in the Greek view, which listed them in order of increasing distance as moon, Mercury, Venus, sun, Mars, Jupiter, and Saturn), it would be fair to consider the farthest planet, Saturn, to be 10 million miles away. The sky itself, with the stars painted upon it, would be immediately beyond Saturn.

A reasonable guess, then, for the shape of the universe in Milton's time would be that of a large sphere about 10 million miles in radius and, therefore 20 million miles in diameter. Such a size would be acceptable to the astronomers of the day whether they thought the earth was at the center or the sun.

If, then, we imagine heaven to lie outside the celestial sphere in one direction and hell outside it in another, we have a vision of three separate universes, each enclosed in a spherical "sky." In book

II, Milton speaks of "this firmament of Hell" so that he must be thinking of hell as having a sky of its own (with planets and stars of its own, I wonder?). Presumably, so would heaven.

Milton doesn't indicate anywhere in the epic just how large he thinks the celestial sphere is, or how large heaven and hell are, or exactly what their spatial relationship to each other is. I suppose that the simplest setup would be to have the three of them arranged in an equilateral triangle so that, center to center, each is 30 million miles from the other two. If all are equal in size and each is 10 million miles in radius, then each is 10 million miles from the other two, firmament to firmament. This is an un-Miltonic picture but at least it is consistent with what he says and with the state of astronomy at the time.

Milton, having postulated three separate universes, each neatly enclosed in a solid, thin curve of metal, called the "firmament," invites the question, What lies outside these three universes?

This question arises in modern science, too, which visualizes our universe as expanding from a small condensed body some 15 billion years ago. What lies beyond the volume to which it has now expanded? ask the questioners.

Scientists might speculate, but they have no answer and it may even be that there will prove no conceivable way in which they can find an answer.

Milton was more fortunate, for he knew the answer.

Later, Milton has Satan point out that the storm is over; that the divine attack which hurled the rebelling angels out of heaven and down a long, long fall into hell has now died away:

"And the thunder, winged with red lightning and impetuous rage, perhaps hath spent his shafts, and ceases now to bellow through the vast and boundless Deep."

In the biblical story of creation, it is stated that, to begin with, "darkness was on the face of the deep." The biblical writers apparently saw the universe at its beginning as a formless waste of waters.

Milton must accept the word, for he cannot deny the Bible, but he grafts onto it Greek notions. The Greeks believed that the universe was originally chaos—that is, "disorder"—with all its fundamental building blocks ("elements") randomly mixed. The Divine Creation, in this view, consisted not in bringing matter into existence out of nothing, but in sorting out those mixed elements and creating cosmos (an ordered universe) out of chaos.

Milton, in the epic, equates the biblical "deep" with the classical "chaos" and has it "boundless."

In other words, in the Miltonic view, God, who is eternal, existed to begin with, but for countless eons was surrounded by an infinite waste of chaos.

At some time, presumably, he created heaven, along with hordes of angels who were given the job of singing the praises of their Creator. When some of the angels got bored with the task and rebelled, God created the companion world of hell and hurled the rebels into it. Immediately thereafter, he created a celestial sphere within which a new experiment might be housed—humanity.

All three, then, are embedded in the still infinite sea of chaos in which, if God chose, innumerable more celestial spheres might be formed, though Milton nowhere says so.

Milton goes on to describe how the fallen angels, in their new home, so different from their old and so much worse, nevertheless get to work to try to make it as livable as possible. "Soon had his crew opened into the hill a spacious wound and digged out ribs of gold."

Although gold is an entirely unsuitable structural metal (too soft and too dense) and is valued only for its beauty and rareness, human beings, completely mistaking a subjective, assigned value for the real thing, have unimaginatively dreamed of golden buildings and golden streets (studded with equally unsuitable precious stones) as the highest form of luxury. They have imagined heaven to consist of such structures, and apparently the fallen angels want to make their new habitation as much like home as possible.

They build a city, which they call All-Demons in a democratic touch that contrasts with the absolute autocracy of heaven. Of course, the name is given in Greek, so that it is Pandemonium. Because all the denizens of hell meet there in conference, the word has entered the English language from Milton's epic to mean the loud, confused noise that would seem characteristic, in our imaginings, of a hellish gathering.

There follows a democratic conference in which Satan, who has rebelled against God's dictatorship, welcomes the views of anyone who wishes to speak. Moloch, the most unreconstructed of the angelic rebels, advocates renewed war: meeting God's weapons with an armory drawn from hell—

"To meet the noise of his Almighty Engine he shall hear Infernal thunder, and for lightning see black fire and horror shot with equal rage among his angels; and his Throne itself mixt with Tartarean sulfur, and strange fire."

The "black fire" is the "darkness visible" of hell. "Strange fire" is an expression from the Bible. Two sons of Aaron burned "strange fire" at the altar and were struck dead in consequence. The Bible doesn't explain what is meant by *strange fire*. One can guess that the unfortunates didn't use the proper ritual in starting the fire or in blessing it.

Here, however, we can't help think in the hindsight of our recent knowledge that infrared is not the only direction in which we can step out of the visible spectrum. There is also ultraviolet at the energetic end, along with X rays and gamma rays. Is Moloch suggesting that the demons counter the lightning with energetic radiation (black fire) and nuclear bombs (strange fire)?

After all, Milton can't be thinking merely of gunpowder when he speaks of strange fire. As is explained later in the epic, the rebelling angels used gunpowder in their first battle and were defeated anyway. So it has to be something beyond gunpowder!

(What a science fiction writer was lost in Milton by virtue of his being born too soon!)

Once the various rebels have spoken, each arguing a different point of view, Satan makes a decision. He is not for outright war, nor for surrendering to defeat either. Suppose, though, that someone were to make his way to the human celestial sphere. There that someone might try to corrupt the freshly created human beings and thus spoil at least part of God's plan.

It would not be an easy task. In the first place, whoever essayed it would have to break through hell's firmament, which "immures us round ninefold, and gates of burning Adamant barred over us prohibit all egress."

Then, even if someone managed to break out, "the void profound of unessential Night receives him next."

This is an amazing line. Consider:

There had been stories of trips from earth to the moon even in ancient times. In 1638, an English clergyman, Francis Godwin, had written *Man in the Moone* about such a trip, and it had been a great success. Milton may well have known about the book, so that the notion of travel between worlds was not absolutely new.

Yet all previous tales of trips to the moon had assumed that air existed everywhere within the celestial sphere. Godwin's heroes had gotten to the moon by hitching wild swans to a chariot and having the birds fly him there.

Milton, however, was talking not about interplanetary travel nor even about interstellar travel. He was speaking of travel from one universe to another and he was the first writer on the subject to realize he would *not* be traveling through air.

The Italian physicist Evangelista Torricelli, by weighing air in 1643, had shown that the atmosphere had to be limited in height and that the space between worlds was a vacuum, but this stunning new concept was for the most part ignored by otherwise imaginative writers for a long time (just as so many of them ignore the speed-of-light limit today).

Milton reached out for the concept, however, when he speaks of a "void profound" and "unessential Night."

Night is the synonym for chaos ("*darkness* brooded over the face of the deep") and "unessential" means "without essence," without the fundamental elements. And yet, as we shall see, Milton reached out but only partly grasped the notion.

Satan, scorning to propose a dangerous task for someone else to fulfill, undertakes the journey himself. He makes his way to the bounds of hell, where he encounters a hag (Sin) and her monstrous son (Death). There he persuades the hag, who holds the key, to open the barrier. Satan then looks out upon the "void profound."

What Satan now sees is a "hoary deep, a dark illimitable ocean without bound, without dimension, where length, breadth, and height, and time and place are lost; where eldest Night and Chaos, ancestors of Nature, hold eternal anarchy, amidst the noise of endless wars, and by confusion stand. For hot, cold, moist, and dry, four champions fierce strive here for mastery, and to battle bring their embryon atoms."

This is not a vacuum that Satan describes, but it is a concept equally daring, for Milton's imaginative description of chaos comes quite close to the modern view of the state of maximum entropy.

If everything is a random mixture and if there are no substantial differences in properties from point to point in space, then there is no way of making any measurement for there is nothing to seize upon as a reference point. Length, breadth, and height, the three spatial dimensions, no longer have meaning. Furthermore, since the flow of time is measured in the direction of increasing entropy, when that entropy has reached its maximum, there is no longer any way of measuring time. Time has no meaning any more than position does: "time and place are lost."

The Greeks divided matter into four elements, each with its characteristic properties. Earth was dry and cold, fire was dry and hot, water was wet and cold, air was wet and hot. In chaos, these properties are thrown into total confusion, and, indeed, maximum entropy is equivalent to total disorder.

Suppose the universe is in a state of maximum entropy, so that (in Greek terms) chaos exists. Once total randomness exists, continuing random shiftings of properties may, after an incredibly long interval (but then, since time doesn't exist at maximum entropy, an incredibly long interval might as well be a split second, for all anyone can say)—chance may just happen to produce order and the universe may begin again. (If a well-shuffled deck of cards is shuffled further, then eventually, all the spades, hearts, clubs, and diamonds may just happen to come back into order.) The role of God would then be to hasten this random event and make it certain.

In describing chaos in Greek terms, Milton, however, does not entirely let go of the notion of a vacuum. If chaos has all matter mixed, there must be fragments of nonmatter mixed into it as well, or it would not be true chaos. Every once in a while, then, Satan might encounter a bit of vacuum, as an airplane may strike a downdraft, or a swimmer an undertow.

Thus, Satan meets "a vast vacuity: all unawares fluttering his pennons vain plumb-down he drops ten thousand fathom deep, and to this hour down had been falling, had not by ill chance the strong rebuff of some tumultuous cloud instinct with fire and nitre hurried him as many miles aloft."

I believe this is the first mention of vacuum between worlds in literature. (To be sure, Milton did not have the notion of gravity straight. He wrote twenty years before Newton's great book on the subject was published.)

Satan makes it. By the end of book II of *Paradise Lost,* he has reached earth, having performed as daring and imaginative a journey as any in modern science fiction.

There's just one other touch, I want to mention. In book VIII, Adam asks the archangel Raphael how the angels make love.

"To whom the Angel with a smile that glowed celestial rosy red, Love's proper hue, answered, 'Let it suffice thee that thou knowest us happy, and without Love no happiness. Whatever pure thou in the body enjoyest (and pure thou wert created) we enjoy in eminence, and obstacle find none of membrane, joint, or limb exclusive bars: Easier than air with air, if Spirits embrace, total they mix, union of Pure with Pure desiring . . ."

When *I* wanted to write about another universe and about a group of living organisms totally different from ourselves, I needed one thing that was really strange about which to build all else.

I had my organisms make love totally and "obstacle find none." I arranged three sexes as a further difference and "total they mix." Out of that came the second section of my novel *The Gods Themselves,* which won a Hugo and a Nebula in 1973 and of which everyone said the second part was best.

So if you want to know where I get my crazy ideas—well sometimes I borrow them from the best science fiction writers I can find —like John Milton.

And if by some chance you suddenly feel interested in reading *Paradise Lost* for yourself, I suggest you find a copy of *Asimov's Annotated "Paradise Lost."* Some people think it's pretty good; *I* think it's extremely good.